Project and Program Turnaround

Project and Program Turnaround

Thomas Pavelko

CRC Press is an imprint of the
Taylor & Francis Group, an **informa** business
AN AUERBACH BOOK

The views presented in this book are those of the author. They are not the views of any of the author's prior employers, of any organization with which the author has done business, or of Lockheed Martin. All outlines and checklists in the figures are copied from open source examples.

CRC Press
Taylor & Francis Group
6000 Broken Sound Parkway NW, Suite 300
Boca Raton, FL 33487-2742

First issued in paperback 2022

ISBN 13: 978-1-03-247703-9 (pbk)
ISBN 13: 978-1-138-62680-5 (hbk)

DOI: 10.1201/9781315212746

Publisher's Note
The publisher has gone to great lengths to ensure the quality of this reprint but points out that some imperfections in the original copies may be apparent.

Visit the Taylor & Francis Web site at
http://www.taylorandfrancis.com

and the CRC Press Web site at
http://www.crcpress.com

Dedication

I wish to dedicate this book to my mother, Adelyn (Stecki) Pavelko, and my father, Francis Pavelko, members of the Greatest Generation, which fought for the safety of the USA. They inspired me to always serve.

Contents

Preface

The U.S. economy thrives on the development of new products, new systems, new processes for organizing and managing people, and new ways of teaching. Usually, these advances come about as a result of a flash of inspiration by highly creative individuals.

However, there is a discrete change in development emphasis when the decision is made to implement their invention. Now, they must take on the grueling details of product capitalization, profit plans, development/manufacturing facilities, talent management, subcontract management, supply line development, development tracking, etc. The leadership culture must realign.

Unfortunately, about one out of every four development programs fail. Sixty-five percent of those that survive do not achieve all their goals.

A development program or project in trouble is distinct from a program encountering typical development difficulties. Each day its performance seems to get worse, even when team members do not think it possible. The program or project can appear to be in free fall.

I spent years at a major aerospace company leading the successful recovery of important development programs in trouble. I experienced firsthand what causes programs to stop achieving major performance, cost, schedule, and quality commitments. I refined ways to evaluate these programs, reorganize them to achieve their commitments, and resurrect the morale of the team. I strove to lead these teams to enter a high state of productivity that I refer to as being "on step." In the years after these programs were saved, former team members enthusiastically volunteered to be assigned to similar programs.

There are other books written about saving troubled programs. But few authors have actual experience planning and leading a program save. Many of these books propose determining the root cause of a failing program from a wide range of theoretical possibilities, when in fact the possible causes are few in number. They often recommend responding with a detailed program plan, when actually planning is not what these programs lack.

This book identifies the essential fundamentals for executing a program turnaround effectively. These fundamentals include assigning responsibility for each program task to one person, capitalizing on colocation and face-to-face communication, recruiting problem solvers, establishing team member commitments, using team accomplishments to propel high team morale, among others. All of these are necessary to get the Program Team on step and successful.

The guidance provided in this book is applicable to all program or project genres, including manufacturing, nonprofit work, education, medicine, investment management, and municipal management.

A special chapter is devoted to dispelling misconceptions about and providing guidance for software development. Software has become a great part of both providing product functionality and assisting with managing product development.

This book is a highly valuable source of insight and guidance for a wide range of readers, including management professionals, students of business, and executives of corporations. Every member of a product or project development team will find its recommendations to be of high value.

If you are a member of such a team, you have been given the special privilege of embarking on a memorable effort that may change the world.

Acknowledgments

I would like to acknowledge the support and patience of the extraordinary leaders and managers in my life. Their examples have highly inspired me and helped me develop my leadership behavior and the success of the programs I have led. They include Vance Coffman, PhD, Danial Tellup, Alfred Smith, Thomas Tadano, Emanuel Dimiceli, Michel Henshaw, Tory Bruno, Anthony Tuffo, Kenneth Dwyer, Julie Sattler, Richard Campbell, Rudy Kuehn, Stanley Hall, Lloyd April, Gregory Caucus, Michel Brieden, Thomas McGrath, Ronald Harten, PhD, Charles Christensen, George Burk, and Professor John Nicholas.

I also would like to thank those who stood beside me and gave me the confidence to move on. They include Helen Danna, Thomas Greenwood, Thomas Dougherty, John Bjeletich, Christopher Thomas, Dr, Ronald Watson, MD, George Eger, John Kowalchik, PhD, Chad Rowe, Gwynn Sparks, Paul Anderson, Al Gegaregian, Barry Baumgardner, Michael Inman, Lonnie Crawford, Mihn Tran, Russel Friedenthal, Thomas Donahue, Mark Waterman, Donald Navarini, Barry Jaynes, Bernadette Ertle, and all the very extraordinary team members of the programs I have had the privilege of leading.

Finally, I want to offer special thanks to John Wyzalek and Ginger Levin, PhD, of Taylor & Francis, Ginger for picking my book proposal and John for leading the entire publishing effort. And a very special thanks to Marje Pollack and Susan Culligan, who both worked tirelessly with Theron Shreve, of DerryField Publishing Services, to complete this book.

About the Author

Thomas Pavelko worked for 37 years for Lockheed Aircraft and Lockheed Martin. He started as an engineer organizing and leading teams to develop embedded computer systems that performed critical flight control and data reduction functions. Eventually, he was promoted to the level of Program Director. He reported to a wide variety of divisions, including Satellites, Missiles, R&D, Electronics, Propulsion, Advanced Astronautics, Commercial Space, Human Spaceflight, and the Skunk Works. During the latter part of his career, he was assigned to assist large commercial and government programs in trouble. For some of these, he became the new Program Manager. All the programs he led were successful.

Chapter One

Great Program! But What's Wrong?

"What happened? When I took this job, the future looked so bright. It's my life's goal to lead a team and be a part of developing something brand new. I was given the job to lead this development program months ago. We were all so thrilled and optimistic when it started. Finally, we could reveal the brilliance of this new idea. I wasn't worried about my personal acclaim but rather the happiness of this exciting new team.

"But now we're stuck. In fact, each day seems to be getting worse! Can this be true? Like were falling into some desperate black hole with no end in sight. We're not achieving any of our promised completion dates. We're overrunning our budget more and more each week. We're even having a terrible time meeting all our basic performance requirements! How long can we continue to ask our customer for relief?

"It seems like when we believe the condition of the program could not be more dire, we fall into deeper trouble. Do we have the wrong team? Did we just somehow miss some big chunks of the planning? Am I somehow blind to what's most important? I'm straining my mind around the clock to figure out what's wrong!

"My team leaders look bewildered. They keep telling me, 'This is just the reality of a new program,' 'It will get better,' 'Every team goes through a challenging time.' But I keep thinking we've exhausted our margins. I can't imagine we'll ever get back on track.

"There seems to be some team sense that it's ok to keep plugging away as we have. I know few team members will come in this weekend or work any extra hours. I think they're trying to tell me to stop getting up tight. The future will eventually work out. 'No need to kill ourselves.' Right?

"My team probably thinks I'm obsessive. One of those 'hyper-achievers' who easily gets up tight. But I've been in this business a long time, and I just can't imagine how any part of this program could be viewed as successful. I thought this job would be a big step forward. It's like I have no idea what to do to make this better. How did I get into this? Are we doomed?"

• • •

Does any of this sound familiar to you? Are you suffering this pain now?

Leading a team of people developing something that has never existed before is one of the most exciting things you can do. Successfully making a brilliant new idea a reality can leave you with a special pride that lasts a lifetime.

All such development programs and projects are tough. In the fire of hope and optimism, they have their setbacks. Sometimes it may feel like the problems are insurmountable. Yet with the tenacious execution of process and persistence, they get through and succeed.

But some development programs or projects fall into deep trouble. They have a discreetly different tone. The bottom seems to fall out of team morale. The program or project cannot achieve any of its performance, schedule, or financial promises. For a long time, each day has been worse than the last. And there is no consistent plan to end it. Team members sense that the failures that are accumulating will not be resolved without a big change.

This book provides insight and guidance for determining the specific deficiencies in a development program in trouble and then putting it back on track. It distinguishes "commitments" from "goals" and shows how to achieve them. It discusses the changes necessary not only in leadership approach and program processes, but also in the self-esteem of the team. It identifies the many long-term collateral benefits to the enterprise by saving the program.

In this book, I define in detail what a development program is. I compare what poor and successful development programs look like. I review the typical symptoms of a program in trouble. I discuss the major reasons I have observed that cause these problems.

I must clarify that, depending on the subject matter being developed and/or the enterprise sponsoring the activity, a "program" may instead be called a "project," often a big project. Either reference is fine. I will use them interchangeably.

Also, please note that I have developed some terminology that I use throughout the book. I have listed these with their definitions in Figure 1.1.

I will then describe how to reorganize and re-plan the troubled program, how to make the transition in the program, and what high-value fundamentals are important during execution—fundamentals that include closely teaming

TERM USED IN BOOK	DEFINITION
TURNAROUND	Saving a program or project in trouble and making it successful.
PROGRAM or PROJECT	Product implementation effort, 30- to 500-member team with potential subcontractors and suppliers. Terms used interchangeably in book. Detailed definition in the text.
ENTERPRISE	The organization, company, or corporation the Turnaround Program may exist in. The organization can be the Program investors and/or business peers for a "startup."
ON STEP	The highest level of execution success and efficiency a program or project may operate at.
TURNAROUND PLAN	The new Program Plan for a program being saved.
CUSTOMER	The individual or agency that has asked for the program or project to be conducted. Usually the source of compensation.
TURNAROUND LEAD	The person who organizes and leads the program or project Turnaround.
TURNAROUND LEADERSHIP	The leaders of the Turnaround program tasks and other program work. They report to the Turnaround Lead.
TURNAROUND COMMITMENT	The first accomplishment the Turnaround must achieve next for the program or project to remain solvent. Often the first of a series of steps to successfully complete the program.
TURNAROUND TEAM	The Program team plus enterprise executives, customer team, subcontractor team(s), and critically important suppliers.

Figure 1.1 Terms and definitions frequently used throughout this book.

with the customer, selecting the best leaders, setting the right priorities and expectations, motivating innovation, encouraging continuous improvement, managing risk, emphasizing the importance of ethics, providing good subcontract management, maintaining accurate progress tracking, establishing the powerful use of metrics, deriving successful corrective action, maintaining scrupulous control of product configuration, knowing what to look for when recruiting talent, and much more.

There is even a chapter on modern software development and how to manage it. All of this is to get the program team to be like a speedboat skimming on top of the water, efficiently reaching its destination. The reader will agree that this excellent state of operation is of the highest benefit to the customer, the team members, and the sponsoring enterprise.

Other books written about saving troubled programs and projects provide a simple paradigm of determining root cause, correcting the deficiencies found, and monitoring the progress of the recovery. This book goes into greater depth to find the common root causes for program failures. This allows the planning of remedies to be more precise and effective.

The reader will learn that the typical causes for program failure are actually few in number. In addition, the reader will see that programs rarely get into trouble because of incompetent staff. Usually the deficiency lies in a confusing program organization, lack of some program planning, low performance expectations from leadership, lack of positive feedback to the team members, and other reasons not related to the competence or commitment of the staff. There is also usually little, if any, change needed in who is assigned to perform the work on the program team. However, what the new task assignments end up becoming, to whom they are assigned, and what the expectations are to accomplish them may change.

1.1 Programs Are Like Speedboats

After participating in and leading a wide range of development programs, I have observed that they operate much like a speedboat (see Figure 1.2).

Imagine you are leaving the dock on this kind of boat. The skipper pushes the throttle forward. You can hear the motor get louder as it provides more power and the boat moves forward. Let's say the skipper applies enough throttle for the engine to produce 10% of its maximum power. In addition, let's say that for this boat this power results in it moving through the water at 10 mph.

As the skipper heads out to open water, he increases the power output to 20%, and you note that the speed is now 15 mph. You're a little disappointed, because you thought that with twice the power you would travel at twice the 10 mph speed, not just 5 mph faster.

Displacement Performance – Good	On Step Performance – Excellent!

• Achieves average cost and schedule commitments. • Product quality and performances promised are reasonable. • Leadership is evaluated as good and consistent. • Organization design and program process seem to have a terminal velocity.	• Achieves product quality and performances at a higher level than many believe achievable. • Issues and failures solved in shorter than expected times. • Consistently delivers ahead of schedule and at lower costs than anyone predicted. • How do they do that?

Figure 1.2 With the right organization and operation, a project or program will achieve a discretely much higher level of performance than will others in all areas of cost, schedule, promised results, and risk management. I have studied this throughout my career. I call this high level of program performance being "on step."

Now you're farther into open water, and the skipper increases the power output to 40%. The boat is going 18 mph. Real big wake, moving big amounts of water aside, and things are getting exciting. But only 3 mph more? Does everyone else on the boat see that we aren't going much faster with all this power being expended?

But now, the skipper nudges the throttle forward so the engine is creating 50% power. Suddenly the boat moves up out of the water and the speed jumps to 30 mph. Then when he applies 70% power the boat is skimming on top of the water at over 50 mph! At these high speeds the wake actually now looks smaller, but you seem to be flying on top of the water! What happened?

With 50% power, the boat discreetly assumed a different mode of performance that allowed it to gain large amounts of speed with the added power. In this mode, the engine power is being used much more efficiently.

In the first mode of operation, the speedboat is simply pushing the water aside as it moves forward. The drag from the water increases quickly with added speed. But our boat is designed so that the water flowing under it causes it to lift up and skim on top of the water when the power is 50% or above. A boat moving in this second mode is said to be "on step." The transition from displacement to being on step is discrete. Being on step requires a minimum power level and the right design.

These two modes of operation for a speedboat apply directly to development programs. Some programs plow through their schedules, sometimes just barely accomplishing their milestones on time. These types of programs are often severely hampered by unexpected adversities.

But then there are programs that somehow perform *better* than their commitments. They deliver early and resolve problems with highly creative solutions. These programs often complete their tasks with unprecedented speed. While doing so, they perturb the rest of the enterprise very little. These programs are on step.

The purpose of this book is to provide guidance to save a development program or large project in trouble and get it on step!

Let's review in more detail what a program on step looks like.

- The program on step achieves or beats all performance, cost, and schedule commitments. It sometimes makes this appear easy. New risks to the program are detected and mitigated early. All team members clearly focus their work on achieving a simple list of promises to the customer. All program estimates have margins built into them, and the use of these margins is tracked through program completion. All program team members are favorably recognized for bringing forth an improvement idea or suggesting a new approach for the program. All improvement suggestions from the team are thoroughly evaluated by program leadership. Team members volunteer to make personal sacrifices to avoid taxing program margins. All team members know exactly what their assignments are and how their work supports the rest of the program. The team members are often found to smile while working. They may be under pressure but rarely look oppressed. There is respect, pride, and trust among the team members.
- Team member confidence is high. There is a high level of cooperation among the different program task teams. The person with the most expertise regarding a particular subject is known to everyone on the program. No one is ever punished for failing with a new solution they believed would help the program. Confidence and pride among the team members are ramped.

- The innovations and new processes developed and applied by the on-step program often set new performance standards for the enterprise. New tools and development methods are carefully introduced by the program to improve productivity, product knowledge, and product quality. The program on step performs quick and systematic evaluations of proposed innovations and improvements. New methods and tools that do not clearly prove to be beneficial are discarded.
- The program on step often exceeds its profit targets. This would seem unlikely, given the many unanticipated issues that occur during the development of something new. Yet often the innovations, process improvements, and team focus of a program on step results in completing the work faster than past program or project performance actuals and experiences would predict. This performance ends up establishing a higher level of demonstrated productivity by the enterprise, which often provides valuable performance actuals for winning future work. A program on step may even result in promoting a higher level of esteem for the enterprise in the business community.
- As mentioned, the program on step often establishes a higher level of documented enterprise performance capability. Its success can greatly increase the estimated value of the enterprise.
- Program team members volunteer to make large sacrifices in their personal time and effort to achieve critical program milestones. They do so because of the importance of what the program is creating and to support the challenging progress their team members are making. Individual contributors take tremendous pride in being known as a member of the program team. Weekends are sacrificed, long days are worked, vacations are postponed, and more. In a program on step, all the team members know and appreciate what must be done to achieve program commitments. There is little need for leadership to communicate program priorities more than once.

What has surprised me is the high level of enthusiasm and nostalgia that team members have for an on-step program long after it has been completed. We all strive to find a professional opportunity that lets us contribute the most value we can. We all want to "make a difference" for something important. This contribution can be in the form of a breakthrough in something like a new pharmaceutical product, a better way to deliver information on the internet, a new spacecraft for humans, or countless other programs or projects. For many of us, the passion to achieve the promise of a new development can even temporarily trump our focus on career promotion, compensation, or job security.

I have interviewed many professionals who were members of past on-step program teams. Most had completed their contribution to this program three to ten years before we met. I observed that:

- Most past members referred to this experience as "one of the best times in my career." Often their role required completing large amounts of work and solving more problems at one time than they and many others would consider normal. This work was often accomplished in a new and rapidly changing program organization for customers who were often very demanding. These past team members acknowledged they had never worked as hard as they did for this on-step program. Yet, they recalled that because they were given exclusive responsibility for completing a specific task, they poured their hearts into providing the highest quality work they could and to deliver it on time. They told me they knew they had all of the program team and leadership ready to step in and get them out of a jam if they made a mistake. They felt confident the team would not punish them or dilute their role. In almost every case, these former team members said their experience on the on-step program was one they would never forget. Many said they would immediately volunteer to be on a similar team if the opportunity became available.
- In fact, many team members from past on-step programs I talked to volunteered to wrap up their current professional commitment or come out of retirement to work on a similar program or project. I was very surprised to witness this. I had seen the hard challenges they had had with their on-step assignments. I have never observed such high a level of sustained program dedication in any other programs or projects in my career.
- Team reunions have been conducted for the on-step program ten years after its completion! These are often organized by the "grass roots" of the team. They often include folks from a wide range of specialties and personalities who were in the trenches together for long times to complete their important work. It is hard to envision a higher complement to the leadership of the on-step program and its sponsoring enterprise.

1.2 How a Development Program Is Defined in This Book

This book discusses how to turn around failing programs that are developing and implementing something new. I provide guidance for getting them back on their committed track.

This includes programs that are providing services or performing steady-state manufacturing. In these cases, they are often implementing a new work model or change of process.

The product being developed does not have to be a completed physical item you can see and hold. For example, it could be a new algorithm for computer software, a detailed design of a new mechanical device or system, a new process design to increase the benefits of a non-profit organization, a new teaching method to increase the math test scores for fifth graders, or a lower cost approach for a growing municipality.

A development program will often implement something that was earlier shown to be achievable with proof-of-concept demonstrations, simulations, or analysis. Development programs bridge the gap between a new concept and its successful implementation. The proof-of-concept or feasibility demonstrations usually only validate the approach. Therefore, creating a detailed design of the new idea is usually a part of the development program. The following adds to this book's definition of a development program:

- Thirty to 500 people on the team (not including customer personnel, subcontract personnel, or vendors). Any smaller and it is often a product team or concept demonstration team with a single task lead. Any bigger and it's usually multiple programs with a different program manager for each one.
- Typical program responsibilities include documenting and maintaining requirements for what is being developed; creating detailed designs and product development plans; managing the plans and tracking work completed to achieve cost, schedule, and performance commitments; and being vigilant for and managing new risks. Program work must include developing and executing plans to integrate and test product elements and establishing a test plan to verify and demonstrate the final product is completed as promised. The program must also select, organize, and manage all subcontractors and vendors. The program must certify that what they have developed is ready for use. The development program's leadership must be the customer's primary point of contact in the sponsoring enterprise.
- The development program may distribute program work from zero to up to thirty subcontractors. If more than thirty subcontractors report to the program, the work might be better managed with more than one program. The subcontractors will usually perform less than 80 percent of the total program work measured by cost.
- Piece parts and supplies are provided to the development program by vendors per supplier contract agreements. These agreements may include

requirements on configuration control, work documentation, root cause determination process, progress reporting times, reporting formants, and more.
- The development program is often the first user of one or more leading-edge technologies, processes, approaches, and systems in the sponsoring enterprise.
- The development program often applies a modified or new format for the Program Plan compared to ones previously used in the enterprise. The output of this program may be distinctly different from anything the sponsoring enterprise has provided in the past.
- The development program might apply a unique Business Plan. The rationale and structure for deriving profit may be new for the sponsoring enterprise.
- As for many new programs, the development program often requires the development and/or application of software. This could include embedded product control software, test software, simulation software, database development and control software, and configuration control software. Fewer and fewer new product developments are free of this dependency. Chapter 15 is a special chapter on managing software development.
- There is one manager responsible for the entire program, called the *Program* or *Project Manager.* When turning around a program from failure, this manager is also referred to as the *Turnaround Lead* or *Turnaround Manager.*
- A Development Program is usually risky! This is the first time the sponsoring enterprise, and possibly any team in history, has attempted what the program is trying to do. All the supporting analysis, simulations, reviews, bench prototypes, etc., may indicate it can be done. But the Development Program must transition these results into something that achieves the anticipated performances and is dependable, supportable, and profitable. Everyone on the program team feels the apprehension of meeting this challenge. They must fight back with success!

1.3 What Does It Look Like When the Development Team Cannot Do Their Best

The most obvious symptom of a development program in trouble is some combination of it falling behind scheduled completion dates, overrunning costs, and/or not achieving the promised features, functions, or performances. Sometimes the team begins to realize that the product performances they promised are unachievable either within the available budget or for some other reason.

Many contracts have provisions for appealing to the customer for relief on promised performances and work. Unfortunately, for a program in trouble, often after their first appeal is made (and presumably agreed upon by the customer), more follow. The development team and customer often become disheartened by this compromise to the original team promise. Some team members may start to feel like they're working on a "mistake." Requests for deviations to the customer can become a way of life for a failing program. This slippery slope is a classic symptom of a development team that is starting to fall apart.

Inability to provide the promised product to the customer is usually the most egregious indication of a team that is failing. But schedule slips and cost overruns may also indicate that a program is in trouble even if the product performances that are eventually delivered are compliant.

Unforeseen problems will occur while developing something that has never existed before. There is absolutely no way to avoid this. An experienced Program Manager/Scheduler will build margin into the schedule to cope with these unforeseen delays. The experience of program leadership, program management, and enterprise executive leadership must be applied to compromise on a margin size that is not unnecessarily expensive yet is sufficient to cover the uncertainties in the proposed work. Too much margin will result in estimates of needed program cost and development time that are noncompetitive.

A simple sum of all the predicted worst-case schedule outcomes will likely lead to a total schedule margin that is too large and unrealistic. On the other hand, in an attempt not to disillusion the customer with a long estimated program duration, program leadership may underestimate the schedule margin needed. A successful program manager who is leading a program on step must derive a well-analyzed margin, tempered with the results of similar experiences they, their team, and their enterprise leaders have had. The results should be supported with empirical past performance data from their enterprise and other enterprises in the industry, if available. If program leadership chooses margins that are too small, the program may appear to be out of control early in its execution.

A major part of preventing a development team from having to dip into their margins is to correct program development anomalies and faults quickly and completely, on the first try. In general, if it takes weeks or months to find the root cause of an anomaly/fault instead of hours or days, if the program tries a series of different fixes for a problem but it keeps coming back, or if there is a growing accumulation of "unverified failures," the program is in trouble. In Chapter 17, I discuss how to determine the true root cause in a timely manner, correct the cause, and then verify that the correction is complete.

Cost overruns are often associated with schedule slips. Time is money, and added schedule time is added cost. However, sometimes a cost overrun can simply be due to an erroneous cost estimate. Mistakes may have been made in the estimates of the required size of the work force, the needed talent mix, future salary increases, cost of supplies, cost of services, etc. The result can be a program cost overrun that continues to grow without a resolution in sight. Such a cost overrun is another symptom of a program that is falling deep into trouble.

Less measurable indicators are apparent when a development program is failing. An experienced program leader will usually detect them when they review a troubled program.

One such indicator is that program team morale is low. Only a few members of the team appear to believe the program will achieve current product, schedule, and cost commitments. Team members might optimistically say that they feel the program will survive, but that it will have to go through one or more "transitions" before it is successfully completed. The team understands that customer and enterprise executives may not be happy with the status of the program, but often there is some sort of rationale within the team that sums up the situation as, "this is the way this business is done," or, "the customer should have known how hard this would be."

Another indicator is that the team is anxious. There is often a palatable sense of gloom over the progress of the program, and many team members are happy to go home at the end of the day. Some may even muse that they are looking forward to accruing enough seniority to retire and "leave the rat race." This failing program environment is hard on the emotions of the team members and may cripple the enthusiasm in the enterprise to engage in future challenging programs.

Many times when a program is in trouble, team members are not provided with the guidance to use their time most effectively. Confused program processes may not provide the team with a clear set of task priorities. As a result, often team members may spread their attention across a set of issues but are not given direction to solve any one of them. Consequently, only a few problems are actually mitigated, adding to the decline of the program.

Without clear priorities and specific milestones, many team members will resort to just putting in their standard work hours. Their expectation may become that eventually somehow their tasks will be completed. Without clear completion requirements and schedule milestones, they have no measurable indicators to judge their performance. As a result, they may have little incentive to make the extra effort to complete their task. A favorable assessment of their work performance seems to be as much about just being present at the workplace as completing important work.

Often, team members in a troubled program assume the habit of performing only the daily work prescribed by their lead. There is little or no team dialog about the quality of work completed or possible improvements to the plans to complete the remaining tasks. Team members and leaders do not grow in their profession, and the program does not adapt quickly to daily changes.

Unfortunately, when team members of a failing program are asked why their progress appears to be slowing, they will often say that it is the fault of some external circumstance or person. With time, it becomes harder for team members to understand that the way their program is organized and being executed is what's at fault.

1.4 Why Did This Happen?

Part of designing a recovery for a failing program is to understand how it got into trouble. This is not an exercise to establish blame. In any program I have turned around or assisted in turning around, the team members and team leadership were performing the best they could. This was unquestionable. Moreover, without exception, the leadership was intelligent, motivated, and highly capable in many areas.

However, managing the transition from having successfully demonstrated proof of a product concept to deploying the product in the marketplace is more complicated and requires more resources than many professionals anticipate. I will discuss some of the typical difficulties faced when making this transition.

1.4.1 Major Change-of-Leadership Discipline

There is dramatic change-of-leadership and team discipline needed when taking a new product idea that has been demonstrated successfully and transitioning it into something that can be sold, supported, and profitable. This high degree of needed change surprises many professionals in the organizations that must accomplish this transition.

The demonstration of new product concepts and prototypes is truly the lifeblood of many companies. Their job is to show that anticipated performances of a new product approach or design can be achieved. It is the critical first step in transitioning a new product concept into a practical commodity. During concept validation, it is usually too early to worry about the details of making the invention something that can be sold and create profits.

But when it is time to make this invention practical and profitable, a new set of parameters must be addressed. These include detailed product performance measurements, operational reliability, fault response and tolerance, product support, safety, product deployment planning, manufacturing, division of work, and much more. These parameters must be addressed and planned for in any detailed business plan that argues for further investment. A reliable estimate of the future investment required by the enterprise must be derived. In addition, the priority of this proposed new product in comparison to the other enterprise products must be established.

It is very easy to mistakenly conclude that the people who invented and demonstrated a product concept are also the best people to lead its implementation. However, the detailed commitments of these two tasks and the talents required are dramatically different. This is a very important lesson to remember! Many brilliant inventors of new products are not interested in the details of implementing their invention. In addition, these inventors are often disinterested in the daily demands, challenges, and stresses of managing large teams.

A product implementation program led by the product inventors often falls short of the change of program culture needed. The program often loses direction, falls behind schedule, and becomes confusing and worrisome to the program team and enterprise management. Only a few professionals can or desire to perform both of these extremely different assignments well.

1.4.2 Staffing Urgency Targets Most Available Personnel

Many enterprises keep some reserve of professionals on standby to support new programs. While waiting for their next assignments, these employees often perform important support tasks that maintain the infrastructure of the enterprise. These tasks can include documenting enterprise work procedures, assisting with small research and development projects, helping with employee training, deploying and managing various professional specialists, assisting with updating enterprise facilities, and more.

Usually these reserve groups have some number of highly rated workers that had successfully completed their last assignments and are waiting for the next. However, when a healthy amount of networking exits in the enterprise, these above-average employees will often receive offers from other programs before they receive a new assignment from the enterprise. As a result, these individuals are usually available for only very short durations between program assignments.

Unfortunately, some of the remaining professionals in reserve may find it hard to find a good match between their interests and the needs of a program. This difficulty can be due to the uniqueness of their specialty; however, it may also be due to the professional's very specific job requirements, which may include how interested they find the work they are offered, with whom they would work, at what location they would work, etc.

While quickly populating a new program, the employees in this second category are often "ordered" to join the program. They may be told that their new assignment may only be temporary but that their expertise is needed to get the program started. This action is often a result of an enthusiastic attempt by leadership to keep the program on schedule.

Unfortunately, this often results in a mismatch of desired talent and/or career expectations between the employee and the program. Programs that have team members who are not happy with their assignment or who cannot perform it well cannot help the program get on step. New programs with even just a few team members in this unfortunate situation can fall behind.

1.4.3 Outstanding History, So Again?

Often a senior/expert enterprise professional has become successful developing a specific solution for an important application in a past program. Their contribution may have been key toward winning and keeping valuable business for the enterprise at that time. This professional may even have received special recognition from the enterprise and customer for their work.

As this expert's assignment ends, it seems sensible to ask them to solve the problems in a new and what may appear to some to be a similar development program. After all, this person demonstrated good working relationships with the past program team and their customer. The solutions this person provided to this program were very successful.

However, this professional's expertise may not completely apply to the new program. One reason may be that the title of the work to be performed sounds similar, but the details of the work are not the same. For example, an expert may have made beneficial improvements to a supply line at an electronics assembly company. But it would be wrong to assume that the same expert will do a great job managing the supply line for a new auto supply shop. Or an expert may have been rewarded for creating a breakthrough high-speed software database. But again it would be wrong to assume that this person will do a great job leading the development of process control software for a cement plant. The words "supply line" and "software" are the same in these two examples, but the substance of the tasks are very different.

The Program may slowly learn that this expert has little real experience that can be carried over to the new task. In addition, the highly valuable solutions they had provided in the past may have become obsolete. The program leadership may begin to recognize that there could be a long learning period for the expert to get up to date before the program receives the same high level of excellence he or she provided before. While the clock is ticking, the expert may become frustrated, the program leadership becomes frustrated, and the program starts to fall behind schedule.

Still another reason this transfer of expertise may not work is that the expert may have succeeded in providing a solution for a different part of the product lifecycle than the new program needs. For example, a successful leader of a team that developed advanced mathematical algorithms for software might be asked to lead a software development team, the mistake being that the capabilities required to code and test software are very different from those needed to develop the mathematics and decision logic implemented with the software.

An expert who receives an inappropriate assignment may become embarrassed, frustrated, and defensive. The experience needed to become proficient in performing this new job will take a long time to acquire. The program slows down and becomes troubled.

As mentioned earlier, many times the expertise such a person brings to a new application is obsolete. New development programs often need to apply improved processes and technologies to a problem. If an expert has garnered a large amount of acclaim for past work, he or she may be reluctant to discard the old approach and may insist on using it in the new program because they are so familiar with it and it brought them recognition before.

For example, a senior expert may insist on applying a successful but outdated hardware-in-the-loop system simulation when modern software can successfully perform the entire simulation at a fraction of the effort and cost. The erroneous application of experts can hamper the progress of a new development program and dilute its competitiveness by proposing out-of-date, expensive solutions.

The high recognition and sense of security that comes from a long-term association with a specific successful solution is hard to leave behind. However,

- **Inventors not interested in implementation details.**
- **Leadership losses focus on most important accomplishment needed next.**
- **Leadership recruits most available team members.**
- **Not targeting "problem solvers" when recruiting.**
- **Successful staff recruited from past projects without relevant experience.**
- **Leadership not familiar with all areas of expertise associated with the program.**
- **Leaders assign task closure to teams, not individuals.**

Figure 1.3 Some of the major mistakes made that cause a program or project to fail, often the result of leadership that has insufficient experience with, or interest in, the breadth of essential concept implementation tasks.

the best experts will differentiate "best practices" from "best solution." They understand that their ultimate value is in their ability attack problems for a new program with an up-to-date understanding of the available solutions. They must have knowledge of the current and most competitive solutions in their field and be able to objectively evaluate and rank them. They must do this while applying their experience-based understanding to develop the best guidance and support for the needs of the new program.

Please refer to Figure 1.3 for the frequently occurring reasons programs fail during implementation.

1.5 Innocent Leadership Mistakes While Trying to Make It Right

When a development program falls into trouble, sometimes one or more typical management actions are taken. Unfortunately, they often do little to improve the program. In fact, sometimes they make it worse. These include the following.

1.5.1 Mandatory Overtime

The leadership of the development program may believe they simply underestimated the size of the job. Or they may believe that there was more cost and time required assembling the new team than they had originally estimated. They surmise that if the team is asked to work a little harder, they will eventually catch up. They believe that after the team catches up, maybe the request for mandatory overtime can be removed. Understandably, program leadership may believe that if they make the overtime mandatory, it will be fair. They may believe that it is justifiable that everyone on the program will have to sacrifice. This mandate may also demonstrate to higher management in the enterprise that the program leadership team can make the "hard decisions."

Unfortunately, this action attempts to turn the natural program organization upside down. When I was just starting to lead small teams, I attended a number of management seminars that showed the ideal program organization chart as being an inversion of the traditional shape we are used to. On these charts, the program manager was at the bottom and his/her authority flowed up through the leadership team to the individual contributors at the top. I thought this might just have been a novel way to surprise us, to get our attention.

But as soon as I started my first assignment as the leader of a development program, I learned that my primary job was different from what I had thought it would be. The upside-down organization chart I had seen at the seminar was correct. I learned that to allow my team to do their very best, I must clearly state what must be done, including cost, schedule, and product performances. Then I must use my creativity, contacts, and authority to make the job of the program team as easy for them as I possibly could. Time after time, when I clarified the critical program needs directly to the program team, they went ahead and planned a faster path to completion than I would have planned. They volunteered to work the swing and graveyard shifts, the seven-day workweeks, the exhausting 12-hour days to meet deadlines and provide the necessary product performances. This, without me asking them to.

Their effort was derived from their own initiative. I realized that no leader could mandate these kinds of sacrifices and expect better results. Instead, the program leader must manage the dedication in the individual team members to achieve the commitments of the program—a job in ways both easier and more difficult than the one I originally thought a program leader had to do. Every good program leader learns that they must clearly point to the next imperative program commitment to achieve, but then minimally tutor the team on how to achieve it.

It became clear to me that whenever overtime is mandated by program leadership, this unfortunately takes away the ownership of achieving program commitments from the team. Leadership loses the high value of the inventiveness the team can derive to get the jobs done efficiently. Team members now end up measuring progress by the amount of work time they put in and not the achievement of milestones for the work needed by the program. One of the essential ingredients necessary to keep a program on step is a complete understanding of the work needed to be completed—and the dedication to complete it—by everyone on the program team.

1.5.2 Leadership Asserting the Responsibility for Program Success Is Solely in the Hands of the Individual Team Members

Often the leadership of a faltering development program is perplexed as to why their program is failing. They may assume that the team is not working hard enough or that maybe the team does not care to do the most important work first. In a mistaken attempt to motivate the team, leadership may tell the team that failure of the program will be on their heads. As unfortunate as taking this action may sound, it may be an understandable reaction to the panic resulting from managing a program that is failing. Leadership may hope that the shock of this statement may motivate the team to improve in some way. This action is usually more likely to be taken when leadership has little experience managing programs.

The fact of course is that the Program Manager is ultimately responsible for selecting, organizing, and orchestrating the work of the program team. The Program Manager is responsible for sizing the program task, choosing staff subcontractor and suppliers, providing clear assignments with realistic expectations, closely monitoring progress, mitigating problems and failures as they occur, adjusting the Program Plan and staff composition when needed, removing road blocks for the team, etc. The program team must know that the success of the program is derived jointly from leadership allocating work assignments effectively and from team members succeeding with their given assignments. In addition, program leadership must provide the needed program infrastructure and remove the obstacles to allow the successful completion of program work. All the working levels of the team are responsible for its success, certainly not just the individual contributors. Sending a message that program failure will be the fault of the individual contributor divides the program team and reflects a serious lack of leadership depth.

1.5.3 Incorporating Higher Levels of Automation

Many times when a program shows signs of trouble, the leadership will look to higher levels of automation, usually computer based, to increase productivity and get back on schedule. Leadership may incorporate new technologies in an attempt to improve, for example, the fidelity of product simulations, the efficiency and accuracy of supply-line management, the speed of software development, the breadth of risk management, and the accuracy and availability of program cost/schedule status. This added automation is often accompanied by one or more new methods for performing the program work. Of course, leadership appreciates that there may be added cost and program delays incurred from procuring these tools and training the program staff to use them, but they usually expect that the improvement to program performance will far exceed the cost of these investments.

Unfortunately, the addition of automation and new methods in this situation are under the duress of demonstrating higher productivity as soon as possible. Program leadership has exposed themselves to the threat of making the program failures even worse by adding this automation because of the pressure to show improvements *now*.

As a result, the automation used is sometimes selected hastily, without solid, empirical evidence of improved performance for similar programs. Team specialists in the program's problem areas will often know that the added automation and methods are unproven. As a result, team morale may begin to degrade at even a faster rate. Team members that are directed to use this automation often feel they are being unnecessarily exercised to learn and perform their work in a different way, without a clear path to improvement. What may have first looked like decisive action by leadership can end up being wasteful and distracting.

1.5.4 Convening a Team "Retreat" or "Offsite"

These events may occur in many forms. They may be conducted for just program leadership, for just the individual contributors, or for the entire program team. Their duration can be from part of a day to multiple days. Some are conducted in a facility a distance from the enterprise workplace. Others may be conducted as close as a spare conference room at the facility at which the program work is being performed.

Usually the purpose of these events is to build trust among the team members and give them the opportunity to jointly analyze anticipated future challenges to the program. In addition, these events attempt to facilitate more objective Program Planning with, it is hoped, a better program outcome, derived from the perspective of a different meeting environment.

The agenda of these events often consists of a series of team exercises that are selected and led by an independent facilitator. This facilitator usually either works for the enterprise or is hired from an outside firm. This expert or facilitator will usually have one or more assistants to help conduct the event. Many times this facilitator will evaluate in detail the existing work interdependencies and temperament of the team members via interviews and questionnaires before the offsite is conducted.

Usually these retreat or offsite events give the attendees the opportunity to speak their minds about processes they do not think are working well. In addition, it gives them the opportunity to speak about other areas they feel need to be improved. It may give the program team the opportunity to more clearly define the functions and responsibilities of each part of the organization and remove ambiguities.

The goal of these events is to improve operational efficiency and increase trust among the program team members. As a bonus, each team member may gain more insight into the personalities and priorities of their fellow team members. It is hoped in turn that this insight will bring the team members closer to each other and that the program will become more productive.

These events usually leave the participants feeling enriched and rejuvenated. They usually have been away with their fellow workers in a different and informal setting and therefore appreciate each other's individuality a little more. It is hoped this enriched perspective will then help the program to operate more effectively in the future.

In addition, these events allow the participants to rest and rejuvenate in a less demanding setting. It is hoped that they will return to the program challenges with renewed energy and enthusiasm.

Unfortunately, the benefits to the program from these events are often temporary. The lessons learned dissipate as the participants re-acclimate to their regular program setting. Actions taken during these meeting are unfortunately rarely completely closed. Even though the program members may have learned more about their fellow workers, they often settle back into their old work patterns. Major changes to save a failing program rarely result from these events.

It becomes clear that the changes to a development program that permanently improve its performance must be established *at its core.* And this core is the program's leadership, the processes that it uses, and the way it is organized.

When saving a development program, the need for urgent action and the need to be open to all improvement ideas cannot be over-emphasized. An appropriate analogy is being on a lifeboat adrift in the open sea.

Surviving on a lifeboat requires the best effort and the willingness by everyone on the boat to do whatever it takes. There can be no consideration of territorial rights. Face-to-face communication must be immediately accessible among all people on the boat. Excellent communication among the occupants is necessary to stay afloat. Every improvement suggestion must be promptly and seriously evaluated. Improvements that are judged helpful must be implemented immediately and checked for completeness. No time can be wasted criticizing an occupant of the boat who has made an error while trying to improve the situation. Conducting a program turnaround with this level of compromise and intense cooperation is essential for the program to get out of trouble and be on step.

1.5.5 Hiring a Motivational Consultant

Some Program Managers may assume that the inability of their program to progress at a fast rate is simply due to low team morale. They may believe that the team members and leadership simply see a daunting task in front of them and need additional encouragement to move ahead.

As a result, sometimes management will hire a motivational speaker to visit the team. Speakers with this gift are exceptional. They usually evaluate and address the fears of the listeners while empathizing with these concerns. They encourage a high feeling of optimism in the team. They may site examples of other teams that beat the odds and pulled themselves out of what appeared to be a hopeless situation. Many times, they will identify the qualities of the teams they have observed that turned their performance around. Sometimes they will assert that the listeners in the troubled program have these same qualities.

Motivational speakers can help energize a team out of a temporary gloom. Sometimes new program teams are intensely vigilant of their performance or overcritical when they start their work, and they therefore evaluate their performance as worse than it really is. Their pessimism may bring their actual performance down. A good motivational speaker might break this cycle of decline sooner than it would on its own.

However, this book in not designed to save the development team's performances by just increasing morale. Instead, it provides the guidance, priorities, and processes necessary to get a development program back on track. The morale of the program team will become intensely optimistic and brilliantly proud when they observe that they are delivering the high-quality results they had promised and are doing so on time.

1.6 Chapter Highlights

- A development program or large project in trouble often degrades quickly:
  - Some staff may think the challenges are typical.
  - Usually each day is worse than the last.
  - Moral implodes.
  - They are distinctly different from programs that are succeeding.
- Development programs/projects are like a speedboat:
  - Two discrete modes of operation.
  - Fastest and most efficient called being "on step."
  - Many benefits of a program on step.
- Definition of a development program or large development project:
  - Transitions proof of concept to implementation.
  - Thirty to 500 team members.
  - Zero to thirty subcontractors plus suppliers.
  - Team commits to often "scary" challenges.
- Details of a development program or large project in trouble:
  - Major shortfalls in schedule, performance, and/or cost commitments.
  - The rate of failing seems to be accelerating.
  - Remaining program risks are not managed.
  - Team members do not know how to use their time most effectively.
  - Growing number of "blame theories."
- Why did this happen?
  - Different leadership interest needed for implementation work.
  - Recruiting the most available employees during start-up.
  - Recruiting poor match of expertise from other programs.

- Innocent mistakes while trying to save the program or large project:
  - Mandatory overtime.
  - Hastily employing new methodologies and/or tools.
  - Telling the individual contributors that they are ultimately responsible for the failure or success of the program.
  - Convening team "retreats" or "off-sights."
  - Hiring motivational consultants.

Chapter Two

Who Leads the Turnaround?

This chapter covers the attributes the leader of a Turnaround Program should have as well as suggestions about where to find such a person. It also discusses the first things this leader should do.

2.1 Who Initiates the Turnaround?

A turnaround is often started when enterprise management decides that one of their programs must be independently evaluated to determine the extent and cause of its poor performance. They may have reached this conclusion independently and/or from input from other members of the program team, including the customer, suppliers, and/or individual program contributors. They realize that changes in one or all of the organization, approach, and processes of the program must be made to get it back on track. They realize that this program is not just enduring the daily challenges that other development programs have, but rather is not achieving contractually required performances, is slipping development milestone dates by large durations, is overrunning budgeted costs by large amounts, or some combination of these, with no resolution being offered.

In fact, occasionally it will be determined to be impossible for the program to achieve all the requirements stated in the original contract. Perhaps the enterprise had mistakenly promised the impossible. In this case, the Turnaround

Program will serve to minimize the residual delivery deficiencies when the program is completed, to avoid program cancellation.

For all successful programs, each program task must be the responsibility of one accountable person, not a team. This person must be responsible for knowing the development status of the task at hand, including adherence to the contract requirements, the cost plan, and the schedule milestones dates. They must have a current knowledge of known task risks, associated mitigation, and risk backup plans.

This accountable person is the final decision authority in the team for all the work done on the task. A team of many may be working the task, but they must report to this one task lead.

In the same way, one person must lead and be responsible for the entire Turnaround Program. In this book, this person will be referred to as the *Turnaround Program Manager* or *Turnaround Lead.*

Again, the request for a turnaround is often initiated by the enterprise executive management if the troubled program resides in an enterprise. Often, this executive is the person to whom the original Program Manager for the troubled program reported.

The request by executive management to take major action to fix a program is usually made swiftly. Most likely the program has fallen into deep trouble by the time the turnaround action has started, and enterprise executives may have classified this program as an enterprise emergency. Enterprise executives are usually prepared to take drastic action to save the program, and they expect dramatic improvement in return. Often by this time these executives will have clarified the status of the program they are aware of to the customer, including what mistakes were made, what the approach is to save the program, and what improvements are expected, including when they will be complete.

At first, when the enterprise executives initiate the Program Turnaround action, they will usually make liberal concessions to fulfill the needs of the action. However, this powerful support can turn into intense scrutiny and painful skepticism if the program does not demonstrate solid evidence of improving soon. No executive wants to be accused of substituting one failed Program Plan with another! They will remind the program team that this is a major problem that must be solved *now.* The rest of the enterprise will sense the urgency. Tighten your seatbelts!

2.2 Qualifications Needed to Lead a Turnaround

Choosing the right person to lead the program save is the most critical decision to make when starting turnaround action. The person chosen must understand how what is being developed works and how to maximize its utility and ability to create profits, while patiently energizing the program team and leading them to focus their work to be the best they can be. These capabilities are much different from those that were needed to invent what is now being developed. The qualifications needed for this lead include the following.

2.2.1 Have Experience as a Program Manager

The Turnaround Lead must have had the experience of successfully leading the completion of a program or large project. This leadership experience should have been for a development program similar to what has been defined in Chapter One.

2.2.2 Have Good Rapport with the Team

The Turnaround Lead must have demonstrated the ability to work well with the members of the development team. He or she must have the ability to communicate with and be respected by all team members, covering the entire range of professional specialties and worker seniorities.

2.2.3 Have a Wide Breadth of Experience

The Turnaround Lead must have had experience supporting many different programs and must have developed an experience-based intuition that quickly detects successful versus unsuccessful program approaches and processes. The Lead should have firsthand experience with program success and failure. They should have experience-based guidelines to evaluate the effectiveness of the different task teams in a program. They should have attempted to apply these guidelines in past programs and experienced their success or failure. They should be able to accurately estimate the ability of each team member to advance the program, regardless of specialty.

2.2.4 Understand the Type of Project or Program They Are Going to Manage

The Turnaround Lead must be familiar with the basic elements of the programs they are going to manage. These include the typical development milestones, cost and schedule tracking processes, and major development milestone events. Familiarity with the customer, or at least the type of customer (e.g., civil government, military government, commercial development, commercial manufacturing, non-profit), is usually necessary. The time needed to save the program will be too short to learn these basics from scratch, and the new Turnaround Lead will have little credibility without demonstrating this familiarity.

2.2.5 Understand the Workings of the Project or Program

The Turnaround Lead must have a broad understanding of the procedural and technical substance in the program they are turning around. It must be deep enough to assess the competence of the program specialist working each program area, as well as the plans they propose. For example, if the Turnaround Lead is saving a large project implementing a new ATM system for a bank, they should understand the elements and flow of all the customer transactions that will be required at each ATM, as well as the electrical, processing, communication, and software requirements; typical operating environments; repair logistics; failure tolerance; etc. If the Turnaround Lead does not have an understanding of the technology associated with some portion of the project/program they are leading, they must have a trusted report that serves as an expert interface to this area. This report must communicate all development and functional performances to the Turnaround Lead so they understand its status. This must include describing faulty planning, deficient performances, and risks.

2.2.6 Tap Team Creativity and Selflessness

The Turnaround Lead must have a high degree of interest and aptitude about the program to be saved in order to inspire all team members to consistently consider how they can better perform their work. These inspired improvements should lead to lower cost, shorter development time, lower deployment risk, more capability, and other stated priorities of the endeavor. At the same time, the lead should inspire the team individuals to work for the benefit of the team, not just themselves. The program or project team members must value that the power of their team is much greater than the sum of the individual efforts. The Turnaround Lead and the program leadership team must constantly set an example for these behaviors.

2.2.7 Be Present and Accessible

The turnaround work will require long, extra work hours by parts of the program team during the program recovery. The Turnaround Lead must be visibly on site when the rest of the team is present. During the heat of program recovery, the Turnaround Lead should minimize business travel to just what is essential. They should delegate travel if possible so they can remain at the program site(s). The physical presence of the Turnaround Lead is critically important to motivate the program team to embark on a new Program Plan and assume an accelerated work tempo.

With politeness and respect, the Turnaround Lead must assume license to walk into any program meeting regarding any subject—particularly meetings focused on critical issues and plans. The Lead should be pleasant and curious and should stand back and listen, observing the tempo and progress of the meeting. If what they see satisfies them after just a few minutes, they can then leave. It should be clear to the meeting members that the Turnaround Lead has not stepped in to take over leadership of the meeting. As a result, the program team will surmise that the Turnaround Lead is fully engaged and is absorbing a firsthand view of front-line progress. The team observes that the Turnaround Lead is always present, interested, and supportive.

If the Turnaround Lead cannot be present at the program work site(s), one delegate must be assigned to represent the Lead. All team members must be informed, by email, program memo, verbal flow down, or other means, when the delegate will be representing the Turnaround Lead. Program team members should be aware that this delegate remains in constant contact with the Turnaround Lead and that all major decisions by the delegate have been made with the Lead's concurrence. Never allow a "team" of delegates to represent the Lead—it must be just one person. This avoids any sense of ambiguity or conflict in the leadership decisions made during the absence of the Turnaround Lead.

When practical, the delegated representative for the Turnaround Lead should be the same person throughout the lifetime of the turnaround. This will minimize any confusion about who is in charge of the program at any given time. Giving each member of an array of developing program leaders the opportunity to be a delegated program lead may be good for leadership mentoring in other circumstances, but this process should not be exercised during the critical time of recovering a program.

The Turnaround Lead must be both accessible and approachable by anyone on the program. While low-seniority team members must be instructed and encouraged by example to work appropriately with the program's chain of command, they should not be prevented from going directly to the Turnaround Lead regarding any question, issue, or concern. Having said this, if the Turnaround Lead is communicating well with the leadership team and has chosen good leaders who are sensitive to the needs and concerns of their reports, there usually will be little need for junior team members to "go to the top."

2.2.8 The Lead Must Be Inspirational and Shrewd

A good Turnaround Lead must greatly appreciate that they are leading a special event. It will not be "business as usual." Their responsibility is immense. Everyone's role on the program is subject to change while the program processes and organization go through any necessary changes for the Turnaround. As mentioned earlier, each team member must be encouraged to think in terms of what it will take to save the program, not just of the near-term gains for their career plans.

2.3 Where Do You Find the Turnaround Lead?

2.3.1 Apply the Original Program Leadership

The first thought may be to enlist the original program leadership to plan and lead the Program Turnaround (see Figure 2.1). This approach may at first appear

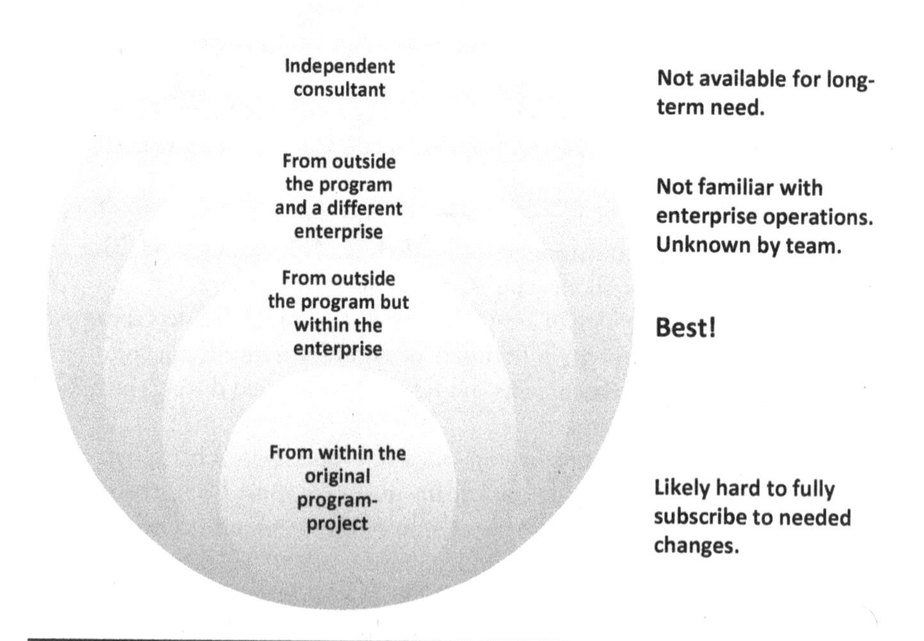

Figure 2.1 There are four basic sources for finding a good Turnaround Lead. One is preferred.

to be the least disruptive and most expeditious—after all, the current leadership is the most familiar with the details of the program organization, the operation of the program, and the individual team members. They should know the strengths and weaknesses of each team member. They should know in detail the history of the program and the status of the specific program elements that are falling behind.

Unfortunately, as mentioned earlier, the original leadership may lack experience and interest in deploying a new concept or product. They have likely been trying to remedy the troubled program with solutions that have not worked. Because of their devotion to their original Program Plan, they may have lost perspective on the critical issues that have since developed in the program's performance.

2.3.2 Recruit the Turnaround Lead from Outside the Program and Enterprise

If a Turnaround Lead has been recruited from outside the enterprise, they will not have a shared history with the existing program team. This history is very helpful to build respect and trust for the leader. If the leader has been enlisted quickly, it may appear to the program team that he or she may leave the enterprise just as quickly sometime in the future.

A new leader without this shared background will often be relegated to the role of a consultant by the team. The new leader can offer no evidence of long-term devotion to the enterprise. They may have had long experience working with the program's customer and an outstanding reputation in the industry in which the program is involved. They may have a special understanding of what the customer perceives as deficiencies in the development program. For some troubled programs and/or enterprises, this may be valuable.

However, this person will *not* have experience with the process details of the enterprise in which the troubled program resides. They will not have a firsthand reputation within the enterprise for the qualities and successes of their work. They will not be familiar with the program team members, including their individual capabilities and professional goals. As mentioned before, they may be perceived by the team as *temporarily* supporting the program, someone who helps provide a fix but moves on before the program is completed.

If the development program is falling behind because of a single deficiency, hiring a temporary Turnaround Lead from outside the enterprise to work with the original Lead may be an appropriate investment. Otherwise, they may well be a distraction during the critical program recovery period and a waste of valuable program resources.

2.3.3 Recruit the Turnaround Lead from Outside the Program but from Within the Enterprise

This person is an employee of the enterprise and often has been for a long time. They usually have a high degree of dedication to the enterprise and show a special dedication to its future interests.

They also usually have a complete and current understanding of the way their enterprise does business. They are familiar with many of the program's team members and may have worked with some of them in the past. They probably are familiar with the history of the program they are being asked to save through the typical internal enterprise communication channels.

If this person has a good reputation for supporting the enterprise's programs and projects, he or she will be immediately respected and trusted by members of the program team. Because this new Lead is an enterprise employee, the program team will surmise that this person will likely stay with the program as long as it takes to turn it around.

Usually a Turnaround Lead from within the enterprise is the best source to lead the correction of a troubled program!

It is even better if the Turnaround Lead assigned from within the enterprise works closely with the original leadership of the program to design the program save. The original leader will have detailed knowledge of the abilities and desires of each team member, as well as the most recent desires of the customer. Also, the program team is used to taking direction from the original leader. Having the original leader work with the Turnaround Lead will make the transition to the new Program Plan occur more quickly and with less embarrassment for everyone.

This new leadership arrangement for the turnaround may be awkward for the original program lead and their leadership team, who will now be working for the Turnaround Lead. However, as discussed earlier, many times this original leader had been erroneously placed in the program leadership role and may become more aware of some deficient performances in their leadership by participating in developing the new Program Plan for the turnaround.

It is even possible that the original program leader can resume their role after the program is successfully put back on track by the Turnaround Program. This of course assumes that the original program leader is truly motivated to learn how to efficiently implement a new process and/or product. It must again be emphasized that this type of work is not attractive to many inventors.

For the Turnaround Lead and the original program leader to work together successfully, the original leadership must fully subscribe to the new Program Plan. Any reservations about the plan must be resolved between the two leaders before the turnaround is started. There can be no second-guessing the new

Program Plan by the original leader or members of their original leadership team after the turnaround work has started. Doing so must be grounds for the Turnaround Lead to take the drastic step of dismissing the original lead and possibly other member of the original leadership team from the program. It is critical that the Turnaround Program team is focused on only one plan.

There can be absolutely NO implication of uncertainty about the credibility or inevitability of a successful program save imparted to the program team members. If the original program leadership remains involved, they must support the Program Plan and execution one hundred percent. The new Program Plan may not be perfect, but establishing a Turnaround Team that works closely together without doubting themselves is paramount.

2.4 First Tasks for the Turnaround Work

First review the basic purpose and status of the program, including commitments (discussed in detail in Chapter 3). Then choose a leadership team and organize it. This team must report directly to the Turnaround Lead. It may consist of elements of the original program leadership team. This would of course be preferred, because team members will already be familiar with the original program's chain of authority and the individuals that performed each role. With existing leaders that are familiar to the program team and trusted by them, less time will be needed for the Turnaround Lead to revise the course of the program.

As will be discussed later in more detail, this is the opportunity to make new leadership assignments from the existing program team members. Many times the expert knowledge that an individual may have from their education, work history, or affiliation with the program is more important than having a long leadership history or refined leadership style. Such individuals can be excellent leaders for executing part of the Program Plan.

Of foremost importance is that each program task be the responsibility of one person and not a team or committee. This description of the assigned responsibility must be absolutely clear to this leader and all the members of the program team. This approach of assigning one lead person to be responsible for each task may seem obvious to the reader but is unfortunately violated in many programs.

A subtler problem may occur when closely associated parts of a task are assigned to different leads, and it can be a mistake to organize this way. For example, the design of a brake disk may be assigned to one team lead and the design of the brake calipers-hydraulic system to another team lead, with no one lead person responsible for the success of the entire brake mechanism. As discussed later, the decomposition of the program work into its pieces must be done such that each piece (one task or part of a product) contains elements that are highly cohesive and independent from the others. In addition, the external interfaces to and from the task must have as little content as possible (often referred to as *minimum coupling*) between tasks. As we will learn, sometimes the customer will have already separated the work into independent tasks as part of their work breakdown description for the job. (We humans intuitively try to break up systems into independent pieces.)

Sometimes a task or product team lead will be asked to be responsible for the management of the personnel of their assigned team, but not for the actual task work being completed or product being developed. By default, the success of the product this team is developing is somehow relegated to the team members. This is a subtle but very serious mistake. The task or product lead must first and foremost be accountable for the complete success of the product their team creates. This lead must be the single point of contact that the other program team members turn to for product status, team planning, and product success. Every member of the program must know this. It must be that simple!

Let me provide an example of how the successful task/product team lead must be responsible for ALL the functions of its product. In some programs there may exist an independent functional team that determines how the interfaces (data, status, control, etc.) among program tasks or functions are defined. Sometimes this functional team will be given the responsibility of defining the interface requirements for these tasks or functions.

This approach is seriously flawed. The lead of any product or function must be responsible for the success of *all* of its functions, including all external interfaces. Each product/task lead must work with the lead of the product/task with which they are interfacing to make sure the interface is thoroughly understood, documented, and works the first time. The leads of each side of the interface must be accountable for the success of the interface, not some other program team. Possibly some functional program group may provide common formats to use in this interface and tools to document the description of the interface, but they are not responsible for its success.

Good task/product leaders should, within the limits prescribed by the Program Manager or any contractual guidance, support the successful use of their task/product after delivery. They should work with the customer to make sure it is used successfully. For many tasks, the job is not really done until the customer is satisfied. It is surprising that some customers will state that they feel a provider was noncompliant, even though their product was delivered on time and was compliant to the contract-specified requirements. Complete customer satisfaction should be a commitment by the program. It is rare that a customer will abuse this principle.

As stated many times in this book, complete ownership of each task or product by one person is mandatory for saving a development program in trouble.

- **Turnaround Lead determines the purpose of the program and its near-term commitments.**
- **Lead determines the next imperative accomplishment needed to keep program alive, called the Turnaround Commitment.**
- **Lead determines status of program.**
    - **Organization**
    - **Meet leadership**
    - **Evaluate work to date**
- **Lead makes leadership changes necessary.**
- **Lead requests each program task lead outline steps needed to achieve the Turnaround Commitment.**
    - **Milestones, durations, and dependencies**
    - **Use professional scheduler(s) to organize**
- **Scheduler(s) integrates all program, subcontractor, suppliers schedules, and customer constraints to draft master schedule.**
- **Turnaround Lead and leadership resolve a challenging plan.**
    - **Achieve the Turnaround Commitment**
    - **Includes margins**
- **Turnaround Lead conducts a program kickoff.**

Figure 2.2 What does the new Turnaround Lead start doing from day one? These are the first steps.

Lack of this individual ownership in the original program may be one of the reasons the program fell into trouble. Figure 2.2 reviews the steps needed to start the turnaround work.

2.5 Customer Involvement When Planning the Turnaround

It is very important that the customer be invited to participate in the planning of the turnaround work. The degree to which a customer wants to be involved in this planning will differ for different customers—but they must be invited.

At a minimum, the Program Plan should be presented to the customer for their feedback, edit, and approval. The customer should be given this opportunity before the plan is formally presented to enterprise executive management for their approval.

At this time, the details of the Program Planning may only be at the top level. Often the Turnaround Team task leads will fill in the planning details after the top-level planning is approved. However, all the leads should have reviewed this top-level plan to make sure it is reasonable (that it accurately addresses all the work to be done and the time required to perform it) before submitting it to the customer and enterprise management for approval.

The Program Plans developed for a turnaround will very in depth but are the new Program Plan for the program. The basic outline for a Program Plan is shown in Figure 2.3.

- **Executive Summary**
  - **Turnaround Commitment statement**
  - **Description of how to reach commitment**
- **Organization**
  - **Organization Chart**
  - **Role and responsibilities**
  - **Change management**
  - **Facilities**
  - **Project completion**
  - **Subcontractors and vendors**
  - **Rules and expectations**
- **Scope Management**
  - **Scope Summary**
  - **Requirements Management**
  - **Configuration Management**
  - **Deliverables**
- **Schedules**
  - **Master Schedule**
  - **Schedule Control**
- **Cost**
  - **Estimation Process**
  - **Budget Allocation**
  - **Budget Control**
- **Quality**
  - **Monitoring**
  - **Control**
- **Human Resources**
  - **Acquisition**
  - **Development/mentoring**
- **Program Interfaces**
  - **Stakeholders**

- Reporting and Communication
- Team interdependencies
- Metrics Collection

- **Risk Management**
- **Procurement**
  - Subcontractors
  - Vendors and services
- **Program Information Management**
- **References (potential)**
  - Integration and Test Plan
  - Quality Plan
  - Safety Plan
  - Headcount Plan
  - Product Support Plan
  - Software Development Plan
  - Risk Management Plan
  - *more*

Figure 2.3 This is a typical outline for a Program or Project Plan. The Turnaround Plan is a new Program Plan and should have the same format. This outline will be referenced later in the book.

In addition, this new Program Plan, called the Turnaround Plan, should include the following information.

- Problem summary
- Program requirements affected by the turnaround, if any
- Résumé for the recommended Turnaround Lead
- Authority given to the Turnaround Lead
- Summary of changes in program or project approach
- Highlights of expected improvements to the program
- Impacts and benefits to the operation of the original program or project, organization, and staff
- Definition of when the Turnaround Program will be considered complete
- Turnaround schedule (new Master Schedule)
- Schedule of planned measurements of recovery
- Schedule of customer status reviews

Sometimes a customer may only want to be informed of program deficiencies found and approve the person selected to lead the Turnaround Program

Plan. Other customers may also require formal approval of the entire Program Plan before proceeding.

It is essential that the customer fully support the Program Plan before approving it. Customer support that is unquestionable will help support the program team during the program changes. A lack of customer support will defeat the turnaround work by diluting its perceived legitimacy as assessed by the program team.

The customer may require a formal presentation of the program's issues, their causes, and the recovery plan before they approve the turnaround work.

All recommendations or directions from the customer *to* the Turnaround Program must be acknowledged and recorded *by* the Program. If the Program cannot implement a customer's recommendation or direction, the customer must agree with the Program's argument against it. Otherwise, the Program must track the implementation of this customer input to closure. The Program must make sure that, if the customer has an official point of contact, this contact fully concurs with the final Turnaround Plan.

Sometimes the customer will recommend specific individuals they believe should be assigned to the Turnaroud Program. These requests should be implemented by the Program if possible. If the Program believes the recommended candidate should not be considered for that role, the customer must agree with the rebuttal. This argument from the Turnaround Program must be supported with solid evidence.

However, the Program should be open to customer-suggested program staff assignments, even if the initial response by the Program to the person being recommended is not favorable. The customer may have observed high performance from this individual in settings the Program or its sponsoring enterprise had not been a part of. Also, there is the added benefit that incorporating customer-recommended program team members will result in the customer's sharing more in the success of the turnaround.

The customer may want to change its own organization or program interfaces, which may include who their counterparts are, so they can better support the design of the Turnaround Team. Turnaround Program leadership must be very flexible in adapting the Program to these changes. They must make any changes that are needed in the Program to support the customer a top priority. The customer must always support and be comfortable with the Turnaround Plan and the way they interface with the Program Team.

It is important that the authors of the Turnaround Plan provide strong evidence that executing it is valuable and necessary. There should be no impression that the Plan is just a way to "try something new." The Plan should contain explicit remedies for program mistakes that have been identified. It should provide credible arguments that the Plan will fix all the deficiencies in the existing program. It should thoroughly explain what original commitments, if any, will not be possible to achieve. The Program Plan must provide a schedule of improvement milestones with metrics that verify that the program recovery is on track.

2.6 Chapter Highlights

- Who calls for the turnaround?
  - Customer
  - Enterprise management
  - Other
- Turnaround leadership qualifications needed
  - Experience managing programs/projects
  - Interest and background in the needs for concept implementation
  - Experience tapping team creativity and process improvement
  - "Jack of all (or many) trades"
  - Being an example and accessible
- Recruit Turnaround Lead from original program leadership
  - Highly familiar with program
  - Must subscribe to needed changes
- Recruit Turnaround Lead from outside the program
  - Best source if still an employee of the enterprise
  - Turnaround leadership need is too long-term for a "consultant"
- What the Turnaround Lead must do first
  - Determine Program status and the next essential program accomplishment needed
  - Develop good program/project organization
  - Assign one responsible lead per task
  - Draft a new Program Plan (Turnaround Plan) including schedule commitments
- Customer involvement
  - Comment on and approve the Program Plan
  - Program must exercise complete process-to-close planning or staffing recommendations from customer

Chapter Three

First! "Point A" and "Point B"

Defining the details of Point A and Point B is the first task the Turnaround Lead must coordinate after selecting the leadership team. This is most often accomplished even before announcing the program organization and completing the new Program Plan (the Turnaround Plan) and staff assignments.

From my experience, it is this knowledge that most failing programs or projects are deficient in (see Figure 3.1). Knowing with a high degree of clarity both

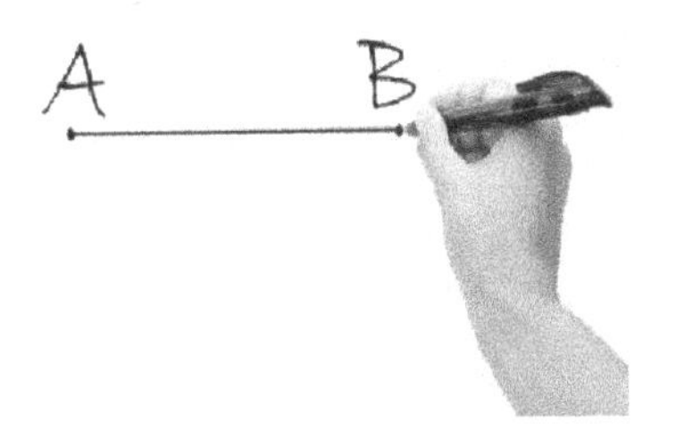

- **Lack of clear definition of these single points often major cause of failing.**
- **Most important information to determine first.**
- **Multiple near-term (point B) commitments are toxic.**
- **Bases for detailed Turnaround planning.**

Figure 3.1 Many programs in trouble suffer from not knowing precisely what they must accomplish next to remain solvent. In addition, they may not have a complete understanding of their status. This information is needed to correct deficiencies and focus future work.

what the program must achieve next to prevent being canceled and precisely what the status of the program is now are the most important pieces of information the program needs to save itself. This information is necessary to plan the shortest and lowest-risk route to program success.

The reader may be perplexed as to how the lack of this information could occur in a team of highly talented professionals and even be the root deficiency for many programs in trouble. How can it be that a team with some of the smartest, most experienced professionals in the world can yet lose perspective on where they are and where they're going?

This problem is often a result of the original program team's having worked program issues at a great depth of detail for a long time. Often the team works so diligently to achieve detailed commitments that they will have either forgotten about, or even rationalized some modification of, what the high-level commitments really are. The program team may not have taken the time to stop, step back, and evaluate the relevance of their work to the next most important commitment to keep the program alive.

I mention now and later that the new Turnaround Lead often has a distinct advantage over the original program leadership when the turnaround activity is being planned. The Lead usually has a fresh, high-level perspective on the program that is not yet encumbered by the many development details.

- **Can be stated in one simple sentence.**
- **Accomplishment needed to keep program/project alive.**
- **Provides sharp team focus.**
- **Consensus of need derived by team members most knowledgeable about product.**
- **Verified as most urgent task by customer.**
- **Is clear and unambiguous.**
- **Will increase the morale of team, customer, and enterprise when accomplished.**
- **Referred to in this book as the Turnaround Commitment.**

Figure 3.2 The next essential accomplishment needed next for the program can usually be stated in a single sentence. It must be a promise, not a goal.

- Original work requirements.
- Updates to work requirements.
- Undocumented customer immediate needs.
- Current program organization.
- Current Program Plan.
- Program Standards.
- Existing Guidelines for the Team Members.
- Team members performing direct versus indirect tasks.
- Current Leadership.
- Current pace of work being performed.
- Metrics accumulated and how they are analyzed.
- State of work completed.
- Location of program facilities and how they are being used.
- Sources of current team members.
- Durations the individual team members have been on the program.
- Who understands the product best?
- The program teams that are strong and weak.
- Team morale.

Figure 3.3 Knowing program status is necessary for developing a credible plan to achieve Point B. It also provides guidance to make corrections in organization and process for the future.

3.1 Details of Point B—The Critical First Step!

Of the two, knowing Point B—what the program must accomplish next to continue to exist—is the first knowledge the Turnaround Lead must have to start planning. With this knowledge, the Turnaround Lead and the program team can start developing and executing a plan to save the program, even before having a complete knowledge of its current status (Point A). Please refer to Figures 3.2 and 3.3.

It is imperative that the program team, while being led by the Turnaround Lead, boil down one single and simple commitment that is imperative for the program to accomplish next for it to be able to continue. This commitment should be amenable to being clearly described in a single sentence.

For example, a commitment that states, "Develop a prototype electric automobile that may be configured to carry four, six, or eight people; be recharged with 120 volts or 240 volts; have a range of 300 to 600 miles depending on a combination of the average ambient temperature and payload weight, within temperature extremes of from 120 degrees to negative 25 degrees Fahrenheit" is too complicated. This commitment must be simplified to what is imperative to be accomplished first. For example, it might instead read, "Demonstrate an electric automobile that travels 100 miles on level terrain without stopping, while cruising at 40 mph, plus or minus 3 mph, and carrying a payload of 720 lbs." This commitment builds a solid first development step without being too aggressive. It is easy for all team members to remember and rally around.

When constructing this statement, the Turnaround Leadership must determine a commitment that is aggressive but achievable and is directly supportive of the development steps that will follow. Keep in mind, this must be a *commitment* and not just a *goal*. A commitment may be less aggressive than a goal, but it is a promise (see Figure 3.4). This planning step will be one the first times to test the ability to integrate the valuable experiences of the new team members.

Please note that in this book I refrain from using the word *goal*. Modern business requires promising what *will* be completed by a specific date, rather than what *might be* achieved at an optimistic date. Program teams rally around a promise, not just a goal.

Many times a commitment includes a smaller scope and states performances values that are lower than a goal would state. However, good customers thrive on providers who clearly describe the work they will complete and who complete it as described, on time, instead of providing optimistic but unrealistic forecasts.

Also, remember that achieving a performance commitment means achieving 100 percent of the commitment. For example, if you promised to provide a system with a reliability of 0.78 over five years, providing 0.7799 over 5 years is noncompliant. The program has not achieved its commitment.

In addition, delivering better performances than required for other parameters in a delivery does not erase the delinquency of a parameter that is noncompliant.

Committing to the work that needs to be accomplished first to save the program in a simple, unambiguous milestone statement has some essential advantages. First, it provides a sharp, single focus for the new Turnaround Team to work toward. No program can be saved without this intense focus on the next critical accomplishment needed. The work needed to achieve this one

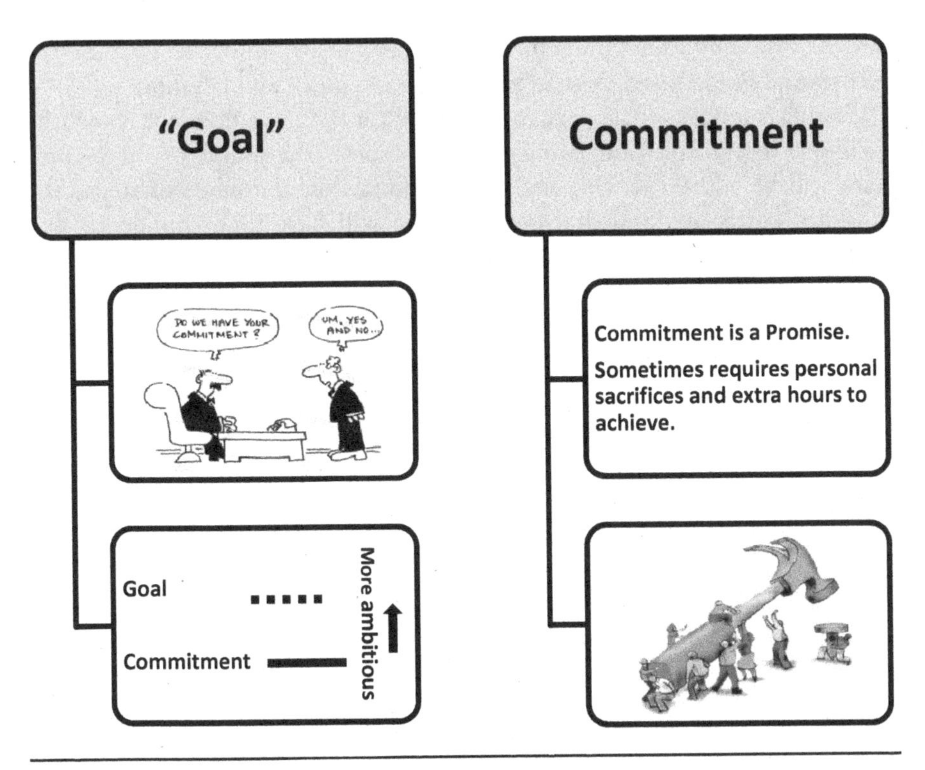

Figure 3.4 The difference between a task *goal* and a task *commitment* is emphasized throughout this book. Any program, especially a turnaround, must maintain "traction" with few surprises. Commitments from the teams are necessary.

commitment may be complicated, but everyone in the program clearly understands what the ultimate outcome must be. Trying to serve multiple program commitments at the same time often leads to confusion and conflict in the program and thus program failure.

Second, this unambiguous milestone is important for team psychology. The satisfaction of being on a team that works closely together and accomplishes major tasks comes when there is a single, simple rallying point for the work being performed. It provides a single and undisputable purpose for the team. All team members know how their success or failure will be evaluated. Differences of opinion in the team are limited to differences in what the best solution is, not differences in what the required outcome needs to be. Multiple commitments at one time for a Turnaround Team often become very demoralizing.

In addition, this clearly stated first commitment by the Turnaround Team is often necessary to keep the program sold to both the customer and the enterprise. When accomplished, it will result in the first evidence of good progress-to-plan. Achieving the commitment will validate the existence of the Turnaround campaign. It is far better for the program to demonstrate the on-time accomplishment of clear and important program development steps from the start of the Turnaround Program rather than to strive for too much and fall short.

Stating a simple, single next commitment for the team does not ignore the likelihood that the program being saved is complex. The complexity of the program will be addressed with the multiple following commitment steps. The customer, enterprise, and development team will have high confidence that the team will fulfill future commitments based on the successes demonstrated achieving the first commitments.

Sometimes the Point B milestone commitment selected by the Turnaround Team must be accomplished to keep the program funded. Many times the enterprise will find a Turnaround Lead if the customer has stated that the program will be cancelled if it does not achieve a specific milestone. This milestone may be stated simply as "completion of validation testing," "first application," "first delivery," "return to positive margins," "reduction to acceptable levels," etc. The customer may have lost patience with the program's poor performance and has issued an ultimatum.

It is essential that the customer accept the Turnaround Program's definition of Point B. The customer may even want to define Point B. Every suggested edit and concern from the customer during its definition must be addressed by the Program and resolved in a way that is acceptable to the Program, the customer, and the enterprise.

The Point B description is referred to in this book as the *Turnaround Commitment.*

3.2 But What Is the Status of the Program *Now* (Point A)?

It is important to wait until the Turnaround Commitment is defined (Point B) before determining the current status of the program (Point A). This may seem counterintuitive, but otherwise, knowing the details of the program's current status, its past experience, and the perceived capabilities of the program team may bias the process of defining Point B. Determining what is needed to save the program must result from a completely independent, objective analysis. Otherwise the Turnaround may fail.

Please refer to Figure 3.3 for a checklist for determining the status of the troubled program before the Turnaround.

It is important for the Turnaround Lead to mingle with and listen to the people doing the program work at their workplace, particularly in program areas that are the most behind. This includes attending team meetings and asking questions in a positive and supportive manner, even offering to assist with solving some of the problems being worked on. The Turnaround Lead must develop a firsthand understanding of the program problems and the people solving them.

The new Turnaround Lead should directly hear why the people doing the work feel the program is failing. The Lead should ask the team how they would organize and execute the work. They should ask them what they think the short- and long-term program commitments should be. The Lead should not leave the impression that they are there to evaluate the individual workers. They are there to learn about the status of the work and build team enthusiasm for identifying the issues and saving the program.

The Turnaround Lead needs to establish informative relationships and trust at all levels of the program. They need to use the information they gather to make accurate decisions for the Program Plan. Presentations from supervisors and managers who review program status will provide an essential higher-level perspective, but the Turnaround Lead must be involved at the worker level as well to make sure the understanding of program status is complete.

The reason the Turnaround Lead is assessing the status of the program is to determine what mismatches exist between the program now and what the Turnaround Commitment will need to be in order to succeed. These mismatches are a guide for promptly making the necessary changes to the program.

There are a number of typical changes that may be needed to make the program successful. Figure 3.5 lists some important changes that are frequently needed.

- **Removal of low gain targets.**
- **Achievable promised performances.**
- **Change of leadership personnel.**
- **Change of organization.**
- **Recruitment of needed talent.**
- **Change of map of work to facilities.**
- **New Program Plan**
  - **New Master schedule.**
  - **Changes in deliverables.**
  - **New work processes.**
  - **New completion definitions.**
  - **New work pace.**
- **Emphasis on commitments, not goals, at all levels.**
- **Change of assigned responsibilities.**

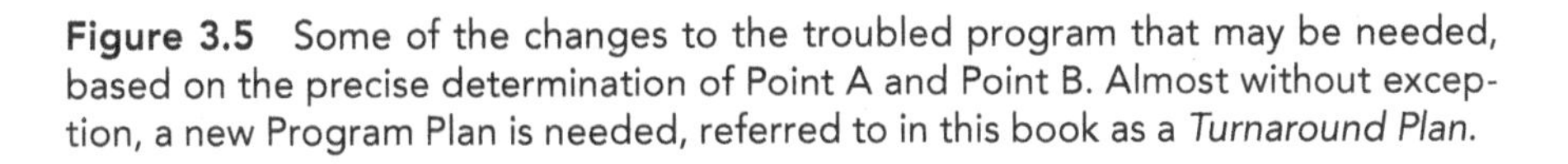

Figure 3.5 Some of the changes to the troubled program that may be needed, based on the precise determination of Point A and Point B. Almost without exception, a new Program Plan is needed, referred to in this book as a *Turnaround Plan*.

Sometimes a team on the program has the necessary number of team members with the right expertise to perform their assignments. However, their work plan does not have the necessary detail to be successful. As a result, this team may not be working on the most urgent part of the program area they are responsible for. The Turnaround Commitment may be in jeopardy.

For example, a program team might be working on their most urgent priority but is using an outdated development methodology. They may be reluctant to increase their work tempo to make up for the slower development pace while using this old methodology. In this case, it will be necessary to update and/or add detail to the team plan to complete the necessary work on time to support the Turnaround Commitment. In addition, the team leadership must remind the team by word and example that extra work hours and personal sacrifices may be necessary to achieve the Turnaround Commitment.

During the planning of the Turnaround, all program team plans need to be reviewed to make sure all work is in support of achieving the Turnaround Commitment. This focus often results in some teams temporarily discontinuing work that is not immediately necessary for the program's survival (which sometimes can have a favorable benefit of reducing the program's total cost to complete!).

By identifying any needed changes to the organization of and/or talent mix of the task teams early in the turnaround work, the Turnaround Program can then get an early start recruiting additional talent. As discussed in detail later, recruiting new talent must be a thorough process to find employees whose work is of high value toward achieving the Turnaround Commitment. For the turnaround work, a department manager, program manager, or professional recruiter cannot just look for some targeted number of "bodies" with a specialty. In addition to a high level of technical ability in their respective areas of expertise, the new Turnaround Team employees should have an entrepreneurial view of the program, including being innovative, being sensitive and interested in increasing profits, being a problem solver, and supporting program success in any way they can. This is true for all the professional areas in the program: manufacturing, engineering, business, finance, management, supply line, quality assurance, etc.

After the new organization for the program has been determined and corrections to the mix and responsibilities of the program staff have been made, it is important that an updated draft Program Plan be created and applied. The time to complete this updated plan should typically be no more than one week. Consider the creation of this plan as a critical act of triage to get the Turnaround Program on firm footing quickly. If there is a delay between the activities of reorganizing the program—including adjusting staffing mix targets and assignments—and starting to create the Turnaround Plan, the credibility of the reorganization will quickly deteriorate. The sense of urgency necessary for the Turnaround Program to be successful can be lost with a long delay between these two actions. Once lost, it will be very difficult to regain.

There will usually be detailed planning needed by the individual program teams/departments following the issuance of the new draft Program Plan. This draft Program Plan (draft Turnaround Plan) will provide a clear structure to build updated team plans that are strictly focused on accomplishing the Turnaround Commitment.

Any errors in the draft Program Plan uncovered by this detailed planning should be corrected as soon as found. These planning errors should be few in number if the initial work estimates provided to the schedulers (to be discussed more later) by the program teams were accurate.

The current copy of the Program Plan must be easy to access by all program team members. Of course, online access is a good way to facilitate this. The entire program team, including the customer, sponsoring enterprise, and subcontractors, should be notified whenever the Program Plan is significantly updated. The magnitude of a change that would trigger a notification should be clearly described in the first version of the new Program Plan.

The program teams may sometimes be asked to commit to new milestones that they are uncomfortable with. Most often these new challenges result in innovations and task focus by these teams that lead to a higher-level schedule performance than they may have thought they were capable of. Setting the bar for the teams at a level that is both challenging and achievable will rely on the judgment of the program leadership.

Between the time the draft Program Plan is done and the detailed planning is completed, a turnaround kickoff meeting should be conducted. It must be arranged and facilitated by the Turnaround Lead and be attended by all program team members reporting to the enterprise, program leadership, and representatives from the enterprise. The customer should be informed of this meeting, but they should be asked to understand that attendance is limited to the employees of the enterprise. Without the customer present, the team members may feel freer to ask candid questions. Subcontractors and vendors may attend as deemed appropriate by the Turnaround Lead. All program team leadership must have approved the new Program Plan (the new Turnaround Plan) and the kickoff agenda before they are presented at the kickoff.

During this kickoff meeting, the program team members should be encouraged to ask questions. The program team will experience a large amount of program change at this time. They should thoroughly understand these changes, review the rationale behind them, and start to feel comfortable with them.

Please refer to Figure 3.6 for detailed kickoff meeting highlights that should be included.

- **Turnaround Lead introduction.**
- **Acknowledge good work performed by the team.**
- **Review of accomplishments to date.**
- **Present Turnaround Commitment.**
- **New Organization.**
- **Responsibilities.**
- **Leadership changes.**
- **New guidelines that may include:**
  - **Commitments not goals.**
  - **Face to face when practical.**
  - **Work measured by completion not time spent.**
  - **Our work needs an extraordinary team.**
  - **Personal sacrifices will be needed.**
  - **On-time delivery of promised work is imperative.**
  - **Enthusiastically open to innovations and continuous improvement.**
  - **Mistakes not good but not a death sentence.**
  - **Root Cause must be determined for development failures.**
  - **All team members must be willing to work any task that is needed now (lifeboat).**
  - **Paranoia is OK.**
- **Near-term milestones.**
- **Q&A.**

Figure 3.6 After a new draft Program Plan (Turnaround Plan) is approved and initial organization changes are made, conduct an all-hands kickoff meeting. These subjects should be included.

3.3 The Customer Must Be Highly Involved

A program turnaround usually results in a dramatic change to the original program. Action is often called for and quickly assembled under the duress of serious cost, schedule, and/or performance deficiencies. Sometimes the customer will be concerned about the validity of the enterprise's quickly making these big changes to the program work plan, even if they are intended to greatly improve program performance. An essential part of a successful program turnaround is to make sure the customer thoroughly understands, believes in, and supports the Turnaround Commitment and new Program Plan.

The Turnaround Commitment and new Program Plan should be introduced to the customer by first reviewing the positive work that had been accomplished by the program prior to the new Program Planning. An analysis showing the current weaknesses in the program and how they will be resolved with the new Program Plan should then be shown. The customer must be thoroughly briefed on what changes are in the new plan, and the Program must obtain the customer's approval of the new plan.

The size of the briefings and other communication to the customer regarding the program changes will depend on the kind of relationship that exists between the customer and the enterprise. For example, if the program-customer primary counterparts are at a high level or if they have established a high mutual trust having possibly worked together closely on past programs, the customer may prefer to stay minimally involved. In this case, the customer may want to simply review the new Program Plan and continue to receive the same periodic reports on program status, with the same agenda and format used originally on the program.

Some customers will choose to closely monitor the progress of their development program. Indeed, sometimes it is the customer who will have requested the turnaround activity for the program. This desire may have been discussed, sometime informally, with the original program manager.

The program may be the customer's first business with the enterprise. The customer may initially view the proposed turnaround action to be risky, distractive, or expensive. In this case, the proposition of program turnaround action to the customer should include empirical evidence of past improvements in other programs resulting from similar actions. In addition, the importance of implementing the turnaround action to maintain the well-being and enthusiasm of the program team must have been evaluated and presented. The customer must understand that low program morale may result in financial loss to the customer for a number of reasons, including lower product quality, cost overruns, and being late to market.

All critique and recommendations from the customer must be accurately recorded, evaluated, and closed as program action items. When addressing these items, compare what the customer says they want and what the enterprise and the program leadership believe the program needs. Often providing well-thought-out program needs will answer multiple customer wants. Determining what issues are common to the needs of both the program and the customer and solving them only once is necessary to use resources efficiently during a Turnaround Program.

Occasionally a customer's desire may seem like more of a personal preference than a necessary need. Do not pay less attention to customer requests that appear to be in this category. Pay the same careful attention to every customer request. Give each one the complete due process provided for in the Program Plan. Steps in this adjudication may include applying change control or failure review processes, holding special executive briefings, and more.

The customer must completely endorse the final new Program Plan. The Turnaround Lead should initiate and conduct any meetings or exchanges of information with the customer that are needed to gain their full support.

Some members of the customer team may have personal reservations about the success of the Turnaround. This should not happen if all customer concerns are gathered, thoroughly recorded, and mitigated during the development of the new Program Plan. No member of the program team should witness any serious criticism or lack of support of the new Program Plan by the customer.

During the reevaluation and re-planning of the program, the Turnaround Lead should appreciate that he or she has a valuable perspective on the troubled program, a perspective which is still based on high-level observed performances, unencumbered by a deeper understanding of the problem details and by closer relationships with the program team members. At this starting point, the Turnaround Lead can change the program organization and team assignments more quickly and with less need to respond to criticism. This less encumbered perspective may become diluted with more time leading the program.

One of the challenges for the Turnaround Lead will be to remain patient with any hesitancy on the part of the original leadership to make program changes. The Turnaround Lead must insist on making the necessary program changes promptly, based on the objective evaluation of the performance data and evidence of past successes from the changes being made. Maintaining completely open and trusting relationships between any original leadership that stays on the program and the Turnaround Lead is highly essential. It will often take patience to maintain these relationships and bring the original program leadership up to speed.

As mentioned earlier, the Turnaround Lead is usually a new member of the program team who has not yet developed professional friendships with team members. Before these relationships begin to develop, it is easier to change individual assignments, add new team members, or find a better program for a team member who is poorly matched to his or her responsibility. The Turnaround Lead is less likely to be asked to justify their decisions during the very start of their leadership role.

3.4 Chapter Highlights

- Point B: What is the major milestone the Program must achieve next to stay alive?
  - Develop a succinct statement called the Turnaround Commitment.
  - "Commitments," not "Goals."
  - Sets up sharp work focus.
  - Provides a target to verify progress and ability to the customer.
- Point A: What is the current program status?
  - Review original promised outcomes.
  - Review task breakdown and program/project organization.
  - Determine progress to date.
- Plan corrections to achieve Turnaround Commitment.
  - Summarize strengths and weaknesses of original plan.
  - Create plan to achieve Turnaround Commitment.
  - Determine new resources needed early to allow time for recruiting.
  - Provide constraints for task planning by task leaders.
- Immediately start triage.
  - All-hands program kickoff ASAP.
  - Track new schedules immediately.
- The Customer must be highly involved.
  - Identify "wants" versus "needs."
  - Present regular status updates.
  - Definitions must be common.
  - Perception is critical.
- New Turnaround Lead has temporary advantages.
  - Fresh "50,000 foot" view.
  - Reassignments are easier.
  - Personnel changes need less justification.

Chapter Four

Find the Cavities

In the last chapter, I discussed the critical importance of the Turnaround Lead's determining why the program is in trouble, what its current status is, and what is most important to accomplish next to keep the program from failing. This critical accomplishment is also called the *Turnaround Commitment.*

Part of the re-plan effort is determining in some detail what is currently wrong or missing that would keep the program from achieving the Turnaround Commitment. Some of the potential causes for this are listed below. I again emphasize how important it is that, once corrective action has been incorporated into the new Program Plan, it be applied promptly.

Figure 4.1 summarizes these typical project "cavities."

4.1 What Hampers Getting to Point B?

4.1.1 Point B Had Not Been Correctly Defined

Sometimes just simplifying and clarifying the next imperative program commitment and focusing the current resources and organization of the program on that commitment is enough to get the program back on track. Maybe to date the program had been promising vague goals but not fulfilling them. Possibly the program was striving to accomplish multiple program milestones simultaneously instead of completing the most important one first. Possibly the stated program commitment was to deliver the final product instead of focusing on the next major interim accomplishment needed to adhere to the Program Plan.

CAVITY	EFFECTS
Wrong Point B.	Wasted resources. Program off track.
Flawed product design.	Successful completion impossible. Design needs update.
Trying the impossible.	Successful completion impossible. Requirements must be changed.
Inadequate workforce.	Product quality will be inadequate/late.
Poor team assignments.	Cost overruns. Low morale. Product quality will be inadequate/late.
Lack of detailed planning.	Cost overrun and schedule slips. Often wrong Point B defined.
Inadequate equipment/ facilities.	Promised product performances not achieved. Cost overrun and schedule slips. Low team morale.
Inadequate subcontract management.	Work not coordinated. Redundant and/or incomplete work. Inadequate achievement of promised product performances. Cost overrun and schedule slips.
Inadequate supply line.	Program cannot achieve promised performances. Poor product quality. Cost overrun and schedule slips.
Progress measured with wrong metrics.	Deteriorating quality undetected in specific product areas. Unable to predict future product test failures. Major cost overruns and schedule slips.
Leaders do not focus on one commitment.	Program or project does not achieve performance, cost, or schedule commitments. Team morale deteriorates.

Figure 4.1 These are the usual cavities in a program or project that cause it to begin to fail. Anyone of these can be catastrophic.

4.1.2 Erroneous Design Concept

Perhaps the proposed design for the product being developed will not achieve its required performances regardless of how much money and effort is spent by the program. It is shameful if the program promised performances that were either impossible or too expensive to achieve, but were knowingly allowed by the enterprise to keep the program sold! I never met a customer who viewed specified product performances as program "goals"—they are instead seen as promises made by the program and enterprise to be delivered exactly as specified.

When the program's design cannot achieve the specified performance, either the deficient design must be changed so it will provide the required performance or the commitment to the customer (often in the contract) must be amended to promise just the contract performances that can be achieved.

Either solution will unfortunately likely result in a sizable expense to the program/customer and possible loss in future competitiveness for the enterprise.

4.1.3 Trying to Break the Laws of Science

Sometimes the contract-required product performances could not be achieved with any design, because the state of the art for the product being developed and/or the limits of the science render these performances unachievable. These errant values may be a result of an unrealistic request by the customer, but the program must take responsibility for signing up for it.

One can question how such a fundamental mistake could have been made. How could a highly capable program team promise to provide an unachievable performance—a mistake that could lead to a catastrophic business outcome for the program?

Perhaps the engineering computations used for this estimate were based on best-on-best case assumptions for all the product elements, as opposed to a statistically normal distribution of performance levels. Perhaps there were assumptions made about future improvements to the product that never materialized.

The only response to this erroneous program commitment is to amend the committed level in the product specification to promise only those product performance amounts that can be achieved. As a result, the development cost to finally achieve these updated performance commitments may greatly increase in value above what was originally budgeted. There might be expensive (and embarrassing) penalties the customer will impose for not conforming to the original performances promised in the contract. Even worse, the customer may terminate the program due to nonconformance or sue for excessive program costs, lost competitive time, etc. They may well take their business elsewhere.

4.1.4 Inadequate Work Force

The workforce for any program consists of an assortment of individuals with special talents. The effectiveness of each of these individuals depends on a combination of the amount of formal education they have had in their specialty, their depth of application experience, the durations of their application experiences, and the interest and drive they have in applying their specialty. This last quality may be hard to measure, yet it is essential for getting a development program back on track.

As will be discussed in more detail later, the work of every program or project must be broken down into its most independent components. Most times the program organization's breakdown into program tasks is based on this work breakdown. In addition, program schedules, funds allocation, cost accounting, and other factors will use this work breakdown to account for all the pieces of the Turnaround Commitment.

It is essential to verify that the right types and amounts of professional talent are allocated to each work breakdown element. This is necessary for achieving fast work completion, efficient use of program costs, and high-quality solutions for the Turnaround Program.

For example, assigning a team of software database engineers to an embedded software application is a poor match of talent to the needed work. Database engineers have little or no experience with developing embedded software. Assigning a team of accountants to lead a cost trade analysis is another example of a poor match. Just because a group of people has a special ability in a portion of a professional area does not mean they can perform all the work tasks in that area well.

Sometimes a program will have people with the appropriate education, experience, and motivation assigned to a task but will have either too few or too many of them (usually, too few).

For example, a development program may have enough test technicians to complete the unit testing of the components of a new product but not enough remaining to support integrated testing. So the program progress is paralyzed later during development, when it encounters an inordinate number of system-level test failures. Or the math scores at a troubled grammar school remain low simply because the school has not employed enough math teachers with the needed training to give adequate attention to all the students.

On the other hand, the program may actually have too many people with the right qualifications working a particular task. This can lead to division of the task into a finer level of granularity than is cost-efficient and necessary, creating a cost overrun for the program. It can also lead to some tasks being performed by more than one program team, which can also result in cost overruns and delays in task completion because of the added cost of coordinating the redundant output of two or more teams.

An example of this waste would be having multiple product quality teams review the same test results. Only one quality team is usually necessary to complete this review. There will likely be a wasted overhead expense coordinating the reviews.

There is a third deficiency that may be found in the program's work force that is harder to quantify but is very important. The program may have, on paper, the appropriate number of staff with the right educational and experience levels assigned to a program task, but the scheduled work progress may still not be achieved. Often the leaders of the under-performing team will offer a complex and circumstantial rationale as to why their team cannot seem to bring closure to their work. Many times the members of this team are frustrated; they may complain about what they perceive as unrealistic expectations for their work performance, and they may blame leadership for not really understanding the challenges in the problems they have been asked to solve.

This kind of deficiency occurs as a result of some or all of the team members having either a lack of experience or a lack of interest in applying their professional specialty. Many times a professional may have an outstanding education but have not yet learned how to efficiently apply their knowledge. Or they are interested in creating development concepts but are not interested in the new work required to implement the concept.

Another distinction that can occur in the program team is that there are usually some professionals who seem to be most comfortable in "problem space," while other professionals in the same specialty who seem to be more comfortable in "solution space." Those in the first group sometimes become anxious when an issue is being closed. They like working and reworking the current problem, refining the definition of the problem, extending the options being considered, and regularly reporting on the work they are doing, but they are reluctant to bring the problem solution to closure. They may feel uncomfortable knowing that after they solve the current problem they will have to come up to speed on a new problem or, even worse, have nothing planned to do next.

But those in the second group tend to be uncomfortable while the program problem remains unsolved. Their greater career joy occurs when they put a problem behind them with a smart solution. This is especially true when their solution uses cost and schedule time more efficiently than planned. The people in this second group are usually more worried about getting the current problem completely solved than about what they are going to do afterwards.

The successful Turnaround Lead needs team members from both groups but with some number of program team members from this second group. It is especially important that workers with this inclination to solve problems as soon as possible be assigned leadership or senior specialist roles in the program. The rest of the program team will look up to these individuals for guidance in the way they approach their individual work tasks.

It may sometimes be necessary for the Turnaround Lead and/or task team leads to supplement or even replace some of the team members in a task group with members that are solution motivated. This action may seem harsh but is sometimes necessary for a program on the brink of failure. Otherwise, the program may continue to languish and burn funds in what is sometimes referred to as "analysis paralysis." It is important that leadership make this correction in the workforce mix as soon as possible. Often a program is saved faster with team members that may only have a medium amount of work experience but have a high motivation to solve problems and move on.

4.1.5 Wrong Task Target(s)

Sometimes, the immediate tasks that have been planned by the original program leadership do not support the Turnaround Commitment. These tasks may be important for the program later on but not now. Any efforts spent on these tasks do not help the Turnaround Team achieve its next necessary milestone to remain supported.

A non-value task may be a remnant from an earlier program plan or may simply not have been correctly scrutinized for value when developed and assigned. An example of this might be funding a finite elements analysis (FEA) of the natural modes of vibration of a mechanical interface, when the failure of this interface to stay intact is known to be caused by the incorrect use of the bonding material. For this example, some may feel the results of the FEA will be valuable latter in the life of the program, so it might as well be completed. But in fact, this activity is robbing valuable resources needed now to keep the program from being cancelled.

When the new Program Plan is being generated, it is imperative that it be scanned for tasks that are not focused on achieving the next essential milestone for the Turnaround Commitment. Often a peer review consisting of both senior program team members and team members newly assigned to the program can scan the new Program Plan and identify low-value work that can either be performed later in the program development schedule or eliminated. As repeated many times, all program resources *must* be devoted to completing tasks that are on the direct-planed path to achieving the Turnaround Commitment.

4.1.6 Lack of Planning Details for Some Tasks

The original Program Plan may lack planning details or may not address the most critical next accomplishment needed to achieve the Turnaround Commitment.

The Program Plan will be broken down to component tasks, some of which may be described with sufficient detail to lead a team. Unfortunately, sometimes in the hurried pace of a new program, other tasks may not be planned with the same detail. When a task team tries to execute a task without adequate planning, the execution details are forced to be created ad hoc. Many times expensive mistakes are made when executing these last-minute plans because they were not generated when the rest of the Program Plan was created. Pressure from short schedule time may not have allowed for thorough review by the program when these plans were generated. Critical interfaces with other tasks may not have been completely thought through and coordinated with the other tasks. With increasing schedule pressures, the program team will be hesitant to stop work and complete the planning of this area. The resulting lack of planning detail will result in expensive misunderstandings of what work is to be completed by which team. In addition, more than one team may end up performing the same work, or important work may have not have been completed by any team when needed by other tasks.

A second mistake that occurs is that the task planning only describes part of the work needed to complete the task. The planning that exists may have adequate detail and be focused directly on the Turnaround Commitment as written, but some necessary portions of the plan are not included.

An example of this mistake could be the description of the task of assembling a computer-controlled mechanical subsystem. The plan might direct in excellent detail the sequence by which the components of the subsystem are assembled. Completing this plan would result in a successfully assembled subsystem. But the Turnaround Commitment requires that this subsystem be completely functional and able to provide certain minimal performances. This requires some amount of functional testing after each step of integration is completed. But if the original plan only addresses assembly without testing the system along the way, it is incomplete.

4.1.7 *Lack of Needed Equipment/Facilities*

Ironically, the Turnaround Lead may have problems acquiring all the facilities and equipment they requested from the enterprise facilities group—this, despite the urgency of saving the program.

This is often true if the Turnaround Program is one of many programs in a large enterprise. This sometimes occurs because the enterprise's facilities team has been encouraged to be balanced and fair in the face of each enterprise program manager's asserting that his or her program is the most important one. So initially, the Turnaround Program is given the same priority as other programs competing for enterprise resources.

But of course, the work of this Turnaround Program is critical. The enterprise has determined that this program recovery is a high priority. If this recovery campaign fails, the finances and reputation of the enterprise may be damaged, or even worse. The Turnaround Lead will usually be given license from enterprise executives to aggressively pursue above-average service from the enterprise facilities group.

The Turnaround Lead must exercise this license. The Turnaround Lead must establish a line of direct communication with the highest level of facilities management. The Lead should rally concurrence from enterprise management to be able to call the facilities manager and insist on highest priority if facilities work is incomplete or behind schedule, or if the facilities plan has to radically change. The Turnaround Lead should diplomatically remind the facilities manager that the Turnaround has the support of the enterprise executive team to turn a key program around. The facilities manager must understand the criticality of this enterprise work to save the program. Sometimes the enterprise executive management will have contacted the facilities management via a separate reporting route, informing them of the criticality of this program save. This further assists the Turnaround Lead in achieving a high priority to use the facilities.

Even with a strong commitment from facilities management, some enterprise facility groups fall behind their schedules. New facility demands are constantly occurring in large enterprises. The facilities organization may be trying to maintain progress by providing the facilities needed to all the requesting enterprise programs. Also, facility organizations are often funded by enterprise overhead and are not directly accountable to profit/loss performance in some enterprises, as are the programs. Therefore, the Turnaround Lead and program leadership must maintain a constant and independent check that the facilities group provides the required facilities and equipment, on time.

In addition to achieving special priority from the facilities organization, the Turnaround Lead must build into the new Program Plan schedule margin for late facilities access. The amount of margin will depend on the facilities provided by the enterprise and how they are applied. If the Turnaround Lead does not have full confidence in the margins they have derived, they should consult experienced program leadership in similar lines of business within the enterprise.

Whenever possible, conduct any meeting with the facilities manager and their team at the Turnaround Program worksite requiring facilities support. This helps reveal issues that may not come to mind in a conference room. Encourage the facilities manager to have representatives from any major outside contractors present at the work site while the Turnaround Program starts to use the facilities. These representatives will help resolve facility issues that may have occurred from facility changes, when the Program starts using the facility.

If a change to the facility plan is made without being coordinated with the Turnaround Program, the Program must report the estimated impact on program cost, schedule, development risk, or product performance to enterprise management. Enterprise management should know about any changes to availability or functionality of facilities that is affecting the performance of the Turnaround Program.

The Turnaround Lead must not allow him or herself to be forced into a role of negotiating enterprise resources with the managers of other programs within the enterprise. Acting as a negotiator will rob valuable attention and time from the Lead. It is the role of the facilities manager to provide the necessary facilities on time. The high level of facilities priority needed to support a successful save of a program may be extraordinary. It may be impossible to derive this high priority from just a balanced compromise between Turnaround Lead and the other program managers in the enterprise.

The importance of having necessary support facilities for the Turnaround Program on time cannot be overstated. Having the needed facilities on time is part of the foundation of the Turnaround. If these facilities are late, or do not function, the momentum of the Turnaround can be lost.

4.1.8 Subcontract Management Is Lacking

Some development programs do a good job of performing their own enterprise work but have problems keeping their subcontracted work under good control. For example, the Program Team may not have a precise or current understanding of the subcontract work that has been accomplished. Their subcontractor may have made changes in their product design or the parts used that the Program is not aware of, or the Program may have only learned about test failures and repairs at the subcontractor through informal channels. The Program may have learned about a test failure or repair a long time after this issue was closed at the subcontractor. Some subcontractors may even go as far as acting like they are leading part of the Program save because of their long history interfacing with the Program's customer and/or having developed many products similar to what the Program needs.

Any of these problems can cause a program to fail. During a Turnaround, all team participants, including subcontractors, must provide current, accurate, and easy-to-comprehend status reports to the single person accountable for the success of the Turnaround—the Turnaround Lead. Subcontract reporting structure and formats are determined by the Turnaround Lead and the Program's subcontractor management team. These requirements must be completely understood, approved, and agreed to by each subcontractor before the Turnaround is started.

Some necessary parts of successful subcontract management are described below. A more detailed discussion of subcontract management will be provided in Chapter 9, Subcontract Success.

Please refer to Figure 4.2.

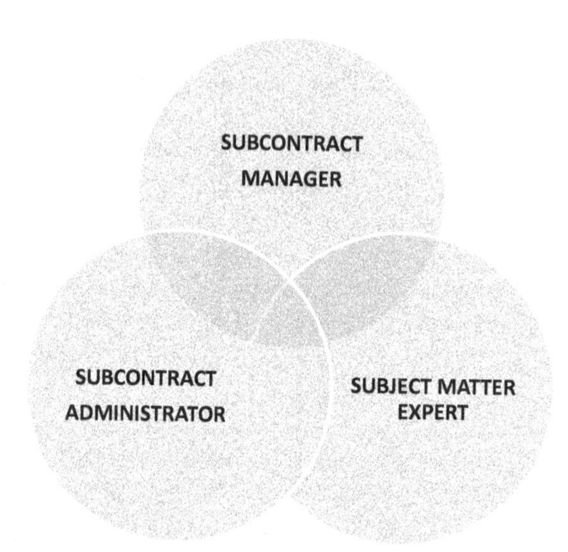

Figure 4.2 This subcontract or supplier contract management organization provides proven management checks and balances.

The Big Three

Each subcontract, big or small, servicing a Turnaround Program must have a Subcontract Manager, a Subcontract Administrator, and a Technical Expert.

The Subcontract Manager leads the subcontract. This person is the point of accountability to the program for the successful delivery of the subcontract work. The Subcontract Administrator and the Technical Expert report to the Subcontract Manager.

The Subcontract Manager usually reports directly to the Turnaround Lead. One Subcontract Manager may manage more than one subcontract at a time as well as be responsible for other program tasks. This of course depends on the size of the workload required for each subcontract.

The Subcontract Administrator usually comes from a functional group of specialists who have been trained on the details of soliciting subcontractors/suppliers, comparing the candidates and selecting a subcontractor, writing a complete and legally binding contract, and coordinating dates and attendance of all meetings and reviews. For a small enterprise or a startup the Subcontract Administrator may have acquired this capability from their past contract experience. In addition, this person verifies the acceptance of all deliverables and the completion of the contract. The Subcontract Administrator usually has been empowered by the enterprise to sign all legal documents and correspondence between the enterprise and the subcontractor.

The role of the Technical Expert is to review and verify that all deliverables received from the subcontractor comply with the requirements specified in the subcontract. Deliverables, of course, cover a wide range, including data searches, trades, research reports, recommendations, simulations, designs, mechanisms, structures, electronics, software, test plans, test reports, and much more. The Technical Expert is often a senior specialist with much experience and success with products similar to that being provided by the subcontractor. The Technical Expert is usually highly respected by the Program and the enterprise for their thorough knowledge of what the subcontractor is providing and for their thorough and fair evaluations.

A Part of the Program Team

The subcontractors should participate in planning the new Program Plan for the Turnaround. There may be proprietary or other data security considerations that moderate the amount of subcontractor participation; otherwise, the subcontractors should be involved as much as allowable. This is especially important if the subcontractors' work is on the Program's critical path toward achieving the Turnaround Commitment.

It is important that the benefits for each subcontractor be negotiated so that each subcontractor is in full support of the Program Plan. Each subcontractor should be enthusiastic about performing their portion of the work and should feel free to recommend innovations to the Program to improve its performance. If the program team senses any skepticism or lack of confidence from a subcontractor, the progress of the program recovery may be diluted.

A subcontractor may have to change their standard or typical subcontractor planning and tempo to support achieving the new Turnaround Commitment. Sometimes, especially for large subcontracts, the subcontractor will assign a special lead to support the increased program pace and changes in deliverables that may be needed in the updated Program Plan. This lead can help to underscore the subcontractor's commitment to achieving the Turnaround Commitment.

The subcontractor schedules must be updated to support the new Turnaround Program Plan with adequate margins. All prime contract and subcontract schedules for the new Program Plan must be in "lock step" with one another. There must be a final audit of program schedules by all program subcontractors and suppliers, prior to the kickoff meeting, to make sure they support all commitment dates and margins in the new Program Plan.

The Turnaround Lead must be vigilant in spotting any wording in any of the subcontracts that allows the subcontractor to charge the Program for any change the subcontractor makes to their support of the program. Charges for such changes must be reasonable and contained. The Turnaround Lead must understand these potential charges in any contract before signing. The Program can allow amendments to the subcontract commitments caused by program changes in the new Program Plan. But these changes must be negotiated between the subcontractor and the Program.

The Turnaround Lead should stipulate for most medium to large subcontracts that their subcontract lead or the lead's representative be present at the assembly site while their product is being integrated. This is so that the subcontract team thoroughly understands any integration issues, is up to date with the integration status, and can provide quick and effective solutions for the program. A turnaround effort can be seriously delayed if a subcontractor needs extra time to understand and solve their own program problems remotely. This can easily force the Turnaround Program team to lose their state of being on step.

Program processes and documentation formats used by the subcontractor should mimic the processes and formats used by the Program. These include design review formats, configuration management processes and definitions, change control process and formats, failure review process and formats, formats and agenda for progress reports, test definitions, test flow, etc. This normalization greatly reduces the number of miscommunications, errors in the use of shared data, and added cost to communicate between the Program and the subcontractor.

The subcontractor must report their development status at periodic program reviews, often just before the Program reports its status to the customer. Either the Program's subcontract management team or the subcontractor's leadership team can report current subcontract status at this review. This report should be in the format used by the Program to minimize confusion and miscommunication. Typically, these program reviews are conducted at least once a week for a turnaround effort.

Often there is a large amount of product development at the subcontractor's facility before they send their work to the Program. For sizable and/or critical work, it is prudent to have a full-time program representative from the Turnaround Program at the subcontractor's facility. This representative can be immensely valuable by providing complete, current, and firsthand status of the development work. Someone onsite can report on subtle development risks that may not be evident in the standard reports from the subcontractor, such as status on present and future adherence to contract milestones, supply-line health, priorities at the subcontractor facilities, and resolutions to test anomalies. Onsite representation is particularly important for subcontractors who provide input that is a major part of the success of the Turnaround Program.

Being a program representative at a subcontractor's facility requires special talents. This individual must understand the subcontractor's leadership structure and flow of responsibilities, their financial mechanism, as well as the technical elements of what is being developed. This person should be respected by the subcontractor's team, demonstrate unquestionable ethics, and be absolutely fair. They must look for ways for the Program to cooperatively work with the subcontractor and avoid blame when problems occur.

At the same time, this representative must promptly and accurately report status to the Program. Often this person will be virtually an ambassador, maintaining a good relationship between the Program and the subcontractor teams while making sure the subcontract work is performed to the standards of the Turnaround Program.

What must not be tolerated is a subcontractor conducting formal or informal discussions with the customer regarding the program without the Turnaround Lead/Program subcontract manager being present. This may occur if the customer has had direct contact with the subcontractor on another program, or if a personal friendship exists between members of the subcontractor team and members of the customer. This kind of interaction must be evaluated as a potential breach of professional ethics, and even the subcontract, if it is encountered. It cannot be allowed to continue.

In rare instances, an unscrupulous subcontractor might independently provide their insight on the Program to the customer or even try to discredit the Program, for their own business gain. This can place the customer in an awkward position and greatly confuse them with biased or false information. Provisions should be written into the subcontract to prevent this. If this kind of interaction is suspected, the Turnaround Lead must make it their first priority to investigate and resolve this suspicion. The customer should be tactfully apprised of any such suspected infraction and informed how it is resolved.

If on-time delivery from a subcontractor is critical to the success of the program, consider a system of payment incentives and disincentives. This is most effective when the terms negotiated with the subcontractor are double-edged. For this type of agreement, the terms should provide an incentive award if the subcontractor delivers the product early, and with all required product checks, reviews, and testing successfully completed. Often this agreement will apply a negotiated formula that determines the early delivery payment amount.

Often the amount of the incentive payment will be based on where the early delivery falls in a predetermined set of early delivery durations. For example, if the subcontractor delivers on time (call it date x) they receive \$$y$. If they deliver between x and $x - 5$ days, they receive \$$y$ + incentive amount. If they deliver between $x - 6$ and $x - 10$ days, they receive \$$y$ + bigger incentive amount, etc.

Late delivery penalties are usually a very effective part of the subcontractor incentive package. This is the other edge of the double-edge incentive approach mentioned above. For some leaders, negotiating these penalties may seem harsh, but they are often more of a motivator than the early delivery incentives. This incentive of course reduces the nominal on-time payment as a function of how late the product delivery is. The late delivery penalty schedule is usually structured the same way as early deliveries for a given subcontract.

Late delivery penalties are very effective, if on-time delivery from a subcontractor is critical. This is because of the basic business perception that adding money to planned revenue is very good, but removing money from the subcontractor's expected revenue is an even a more serious infraction. The consequences to the subcontractor's managers of having their subcontract penalized due to late deliveries are often harsh.

Often the total cost for a subcontract to deliver product with an incentive plus disincentive structure ends up being more than without this structure. However, it may be an excellent investment if the subcontract deliveries are on the Turnaround Program's critical schedule path for success.

4.1.9 Inadequate Supply Line

Many times the purchase of services and products for a program do not require a subcontract. A subcontract usually includes a statement of work, product specification, development schedules, and other pertinent components. But in this case, you are not asking a supplier to develop something. Instead, you want to purchase an existing product that a supplier offers that is often advertised in a product catalog, often referred to as "off the shelf." In this case, you normally only need to generate a purchase order that usually includes a description of what you want to buy, part number(s), how many of these items you want to purchase, and an estimate of what the purchase will cost.

Examples of these kinds of items include electronic piece parts, mechanical fasteners, legal services, accounting services, facility rentals, office equipment, etc. Some of these items, such as metal stock or report binders, may be directly integrated into the product and/or service being provided by the Program. Other items, such as office computers and utility costs, are used by the enterprise to create the goods and/or services the enterprise is delivering but are not actually used in the delivered product. The vast majority of issues with purchase order goods that threaten the success of a development program have to do with the first category—those items incorporated within the product being provided.

Well-executed programs must have quality checks and reports from vendors providing products via purchase order. These reports should include the quality metrics for the product being provided.

Usually the trends of the quality of the vendor's products over time will be evident in these reports. It is important that the program have a review process in place to identify and track any quality trends that may forecast product failures. For larger programs, a quality team can do these reviews. For smaller programs, this review is often performed by a combination of team members from the group using the product and the purchasing team.

One of the important metrics to monitor is the rate of failure during final acceptance testing of the product being provided. The vendor may have a certain limit of acceptance test failures that they deem acceptable. It may be agreed that a failure rate above this limit would trigger a negotiated corrective action plan. This limit is often normalized as failures per delivered item. The number of failures per item may be below this limit at a given time. But the astute Turnaround Program will monitor these statistics for trends and take action if the failure rate is increasing, before it hits the acceptance limit. This kind of trend monitoring of metrics will be discussed in more detail in Chapter 8.

The option to "take action" must be established and defined in the Terms & Conditions section of the purchase order. This wording may include the option to stop delivery and payment.

A frequent mistake made by development programs is to underestimate the importance of monitoring the quality metrics of components and products being provided. These include both subcontract-provided and many off-the-shelf components. All deliveries must be monitored for adherence to clear and fixed quality limits, shown by testing as written into the subcontract or purchase order.

I have had numerous leadership experiences in which degraded quality values or quality trends for relatively simple parts have placed a development program in serious jeopardy. The root causes for these instances of declining quality included errors in assembly procedure, undisclosed change of part vendor, undisclosed changes in assembly procedures, and changes in assembly personnel.

Beyond monitoring quality trends, the Turnaround Program must insist that some negotiated version of a failure review board (FRB) be conducted for every acceptance test failure conducted by a subcontractor or supplier. A report including findings of root cause and corrective action should be provided within some maximum time from the occurrence of the failure, agreed upon in the purchase agreement/contract. It is very beneficial to have all the supplier FRBs follow the same process and develop the same documentation artifacts as used by the Program. This makes it easier and faster to communicate the description of a failure and the corrective action to the Program and the customer. This also results in less misunderstanding of failure cause and correction among all parties. Again, these requirements for standard formats and process must be written into the Terms & Conditions as part of the purchase order or subcontract.

This recommended large amount of failure/correction reporting may seem like excessive oversight and expense for some programs. Yet, quality anomalies that escape the awareness of the Program are very costly and can paralyze development progress. Sometimes just poor management of supplier and subcontractor quality can be the source of cost and schedule overruns, causing the need for a program turnaround. It is very important that the failure monitoring and correction reporting recommended above be required in all program subcontracts and supplier agreements when saving a program. If these conditions were not written into the original program purchase agreements, negotiations must be attempted to add them.

Another major issue that often hampers the progress of a Turnaround Program is the obsolescence of a part manufactured by only one source. The program engineers developing a new design might call for a part that has been manufactured for a long time, usually because this part worked well for the engineering team in past applications. Unless recent enhancements to the performance of this part made by the manufacturer will benefit the Turnaround Program, there may be no need for the Program to spend the time and money to engineer the application of an available newer version of the part.

However, there is often a resulting tradeoff between the impact on cost and schedule of incorporating an updated part versus paying the supplier a premium to continue to manufacture the old part. The premium may be very large, and some suppliers will even refuse to manufacture the old part, no matter how large a premium is offered.

If the supplier states they will not continue to manufacture a part in the future, make sure there are enough spares of these parts procured to cover any failures of the part during product development and application. Also, be careful that the same quality standards and checks are applied by the supplier of this part until its production is completed.

As a last resort, sometimes the obsolete part can be found via a web search. Many times a company will have excess inventory of a part they thought they would need but no longer do. These companies are usually very motivated to sell this excess inventory. Many times parts bought this way can be procured at a large discount. Of course, the source, age, history, storage conditions, etc. of these parts must be determined and thoroughly evaluated by the Program to assure parts quality.

Finally, a mistake that often occurs when a part is being purchased is that the Program does not appreciate that the unit cost will usually change significantly with the size of the order. As a result, if the program decides to reduce the size of an order, the total cost of the order may be reduced less than the program may have predicted. This may have a negative impact on estimated program cost savings from ordering fewer parts. Program leadership must thoroughly understand and plan for a possible increase in the cost per part if the size of their order is reduced.

4.1.10 Program Progress Measured by Work Time Spent Instead of Tasks Completed

Some program leadership cultures give high praise and credit primarily for extra hours worked and personal sacrifices made by the team. Unfortunately, this praise may be given without sufficient regard for the work actually accomplished to achieve the Turnaround Commitment. This may result in what appears to be numerous heroic efforts by program team members, but with little progress toward completing the essential tasks of the Program. The resulting low amount of work completed may in fact be one of the reasons the development program is in trouble.

Please recall as described in Chapter 3 that all Turnaround Team members must commit to completing a clearly defined amount of work at a specific time. This discipline is one of the essential ingredients for a program to successfully execute the new Program Plan (maintain "traction") and stay on step (see Figure 4.3).

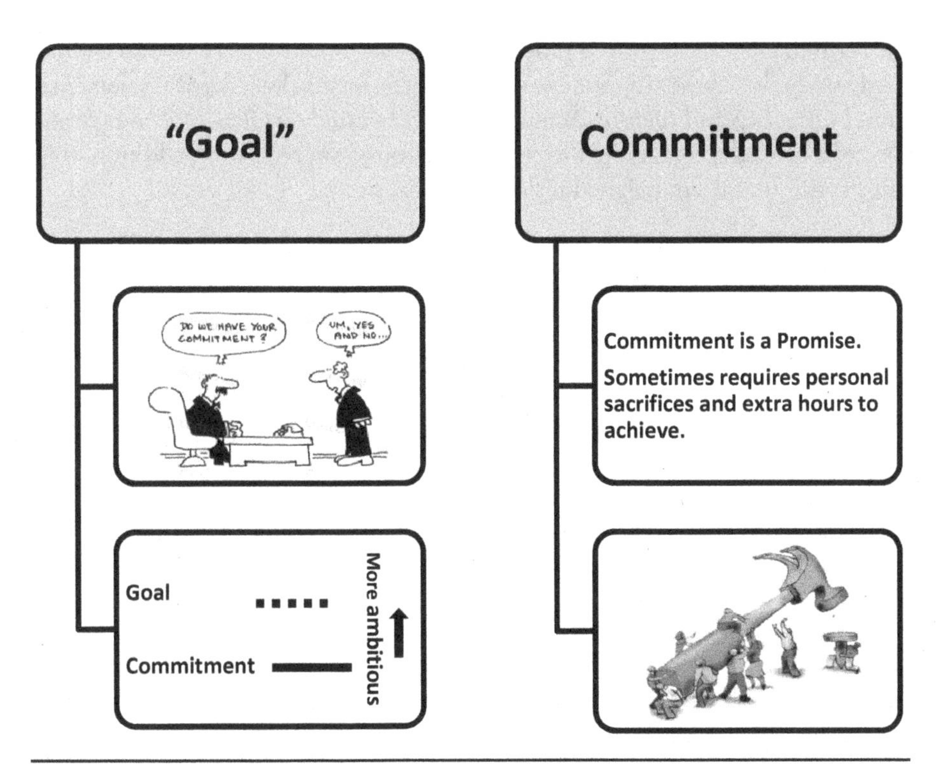

Figure 4.3 Distinguishing commitments (promises) from goals for all program team members is critical. A commitment can have margins built in but is expected to be honored, even if extended work hours are required.

The Turnaround Lead must determine if the time spent by the program team is being used efficiently to achieve the Turnaround Commitment. They may ask if the team is working on the most important tasks to achieve the Turnaround Commitment. Are there trade studies, analysis, development, or other work being performed that do not support the Turnaround Commitment? If so, how should the Turnaround Lead change the Program Plan?

Sometimes the program team's effort is in fact all directed toward achieving the Turnaround Commitment, but the methods and/or processes they are using are inefficient or inaccurate. As an example, the software design for the program may have been put together without using a software design process. Or the evaluation of student math competency may have been collected directly from the individual teachers in a school, instead of through via an independent evaluation.

As mentioned throughout this book, one of the most important principles that applies to a program is that there must be *only one lead* assigned to each program task. This lead must emphasis and lead continuous process improvements for the task to maximize teamwork efficiency. This lead must encourage the team to look for improved processes, methods, and innovations and competently trade the savings in money and time spent versus the risks of implementation. Leadership and their team must be solely focused on efficiently completing the tasks needed to make the Program Turnaround successful. In addition, they must verify that their assigned tasks are complete.

4.1.11 Leadership Did Not Understand the Big Program Picture

Sometimes some of the program leadership team may not appreciate the urgency of their circumstances. They may not understand how much trouble the program is in. Without correcting this deficiency, the new Turnaround Lead will find it impossible to recover the program.

There are a number of possible causes of this inadequacy, the most common of which is a lack of management experience with product development. The program leaders may not appreciate that taking ownership of a program task includes making sure that all the parts of the program that interface with their task are on time. They must make sure neighboring tasks are completed with the same high standards of quality they are applying in their task. This requires that all program leads keep informed of the "big picture" of program status, including the health of neighboring tasks.

If a task leader does not understand the needed leadership scope and task closure that is part of their role, they will usually not significantly acquire this after the program save is started. There is usually not enough time during the rapid initiation of a turnaround program to train a leader to effectively manage work closure. It can be taught, but it takes time and multiple experiences. Also, the task lead must be interested in adjusting the scope of their responsibility and working at often a higher pace. Unfortunately, leaders that are deficient in these ways often need to be reassigned to allow the Turnaround Program to succeed.

Sometimes leaders from the original program thoroughly understand the higher work pace and highly cooperative task interface work they must perform to recover the program, but for personal reasons have decided to not make this extra effort. If so, their decision must be respected. If there are other opportunities in the enterprise that support their preferred level of work intensity, they should be given the opportunity to pursue these. Being on a team that turns a problem program around takes an extraordinary amount of professional effort. Not everyone can afford this high effort.

As mentioned earlier, those ingenious individuals that created a valuable concept or system are many times not interested in developing and implementing it. They may not have the interest or experience to exercise the rigors of detailed program planning and execution. They may not have the interest to lead larger teams with members having widely different backgrounds and personalities. These professionals might instead want to help develop the next big product concept.

When professionals like this are placed into leadership roles for a development program, they often have no interest in understanding the "big picture" of the program implementation. This is a perfectly natural reaction by some inventors and must be highly respected. If the Turnaround Lead finds they have task leads like this that are discontent, they should help the enterprise find a new product concept development position for them.

4.2 Chapter Highlights

- Typical mistake fixed with the Program Plan
  - Wrong Point B
  - Flawed product design
  - Trying to create the impossible
  - Inadequate work force
  - Poor team assignments
  - Lack of detailed planning
  - Inadequate access to needed equipment/facilities
  - Inadequate subcontract management

    - The basic management team of three
    - Subcontractors as part of the program team
- Inadequate supply line
- Progress measured with wrong metrics
- Leaders with little experience focusing on one commitment
- Leadership not planning for all the program work

Chapter Five

Change Gears Now

A big challenge for the new Turnaround Lead is to quickly lead the development of the new Program Plan (the Turnaround Plan) and start executing it immediately. Usually, all program participants, including the program team, program leadership, and potentially, vendors, subcontractors, enterprise executive management, and customer, will see the turnaround challenge as necessary but arduous. These participants understand that there will be risks, process readjustments, and hard efforts ahead to get the program back on track.

There will be a natural tendency by some program team members to make the necessary program changes at a comfortable pace and possibly to even postpone important milestones. Yet, to be successful, it is imperative that all the necessary changes be made quickly. This will usually require respectful but relentless pushing by the Turnaround Lead and the new program leadership team to initiate the new Program Plan as soon as possible and to promptly assume its new pace.

Please refer to Figure 5.1.

5.1 Evaluate Past Deficiencies Quickly

The previous chapters provided guidance on how to distill and focus on the program goals and remedy the deficiencies that caused the original program to start to fail. This analysis must be concluded quickly.

A good target duration for completing the re-planning and embarking on the new schedule is seven days or less. This may seem short to some, but keep

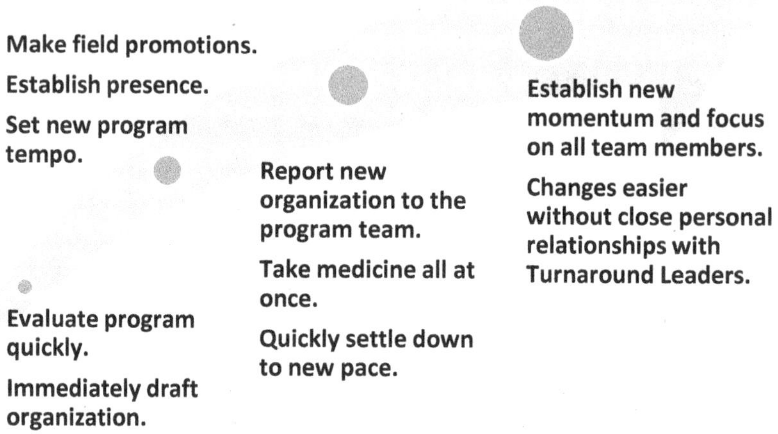

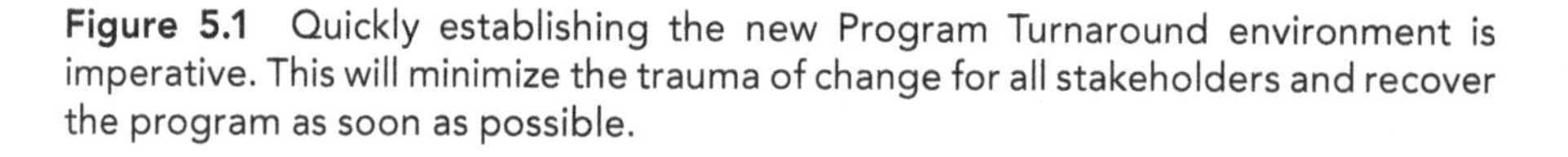

Figure 5.1 Quickly establishing the new Program Turnaround environment is imperative. This will minimize the trauma of change for all stakeholders and recover the program as soon as possible.

in mind that detailed planning can be concluded later. In fact, early detailed planning may be wasted if needed changes to top-level planning are discovered as the execution of the new Program Plan is getting underway.

For an example, in one large and complex program turnaround I led, the leadership team took just three days to build a new Program Plan to save the program. Turnaround analysis and planning must be the highest priority, as this plan will be necessary for the Program to survive. But every day spent in planning takes away from the progress of achieving the Turnaround Commitment.

5.2 Fill The Cavities First!

There will be some refinements and updates to the Program Plan after it is first issued, but usually not many. The most important work to complete first is to assemble a new Program Plan that corrects the deficiencies discovered when determining what was preventing the program from achieving its original commitments.

For example, if achieving low-cost manufacturing has been deficient and is part of the Turnaround Commitment, the Turnaround Plan might require members of the manufacturing group attend the program design meetings to verify that the new designs can now be built at a low cost. If the Turnaround Commitment requires delivering a fully functional product and not just an assembled product, then the new Program Plan might require that unit and integrated functional testing must be a part of the Turnaround Commitment.

Every program deficiency found must be addressed in the new Program Plan. Not fixing one of these deficiencies can lead to continued failure of the development program.

5.3 Draft the New Organization Immediately

This is where the program teams will first see any addition of new program leaders and/or reassignment of others. These changes may be disconcerting for some team members from the original program, so they must be made fairly, discreetly, but quickly, allowing the work to start progress on the new plan without time wasted by any "second guessing."

In addition, this will be the first time the program membership sees the updated reporting structure. It could show original program leaders who are retained but with different people reporting to them. It may show leadership positions that remain unchanged but with new leaders performing this work. Some portions of the organization may assume a simpler structure while others may become more complex.

The new Program Plan and organization will include an updated estimate of the required number of employees having each specific talent. Any new specialties needed on the program team will be identified.

5.4 Making Field Promotions

When choosing different or additional leaders for the Turnaround, the best candidates are often those people currently working on the program.

As part of many management-training curricula, a distinction is made between team *management* and team *leadership*. Team management is generally referred to as the diligent accounting and leadership of program expenditures, schedule status, planned work to go, compliance to contract requirements, risk management work, etc.

Team leadership, however, is derived from the leaders' special knowledge and experience in the program subject area, allowing them to forge the most expeditious path toward achieving all contract requirements. Good leaders consistently encourage their team to offer ways of doing their work with fewer resources, less risk, and higher-quality results. Their door is always open to new ideas.

Often a management class will discuss a "four square" diagram with every combination of good and bad management and good and bad leadership demonstrated by a team lead. They review how the results of analysis show that the best program leaders are those who excel at both management and leadership. However, those who are good at only leadership are very close to having the same high level of effectiveness.

In a Turnaround, team members who understand the details of the problems they must solve and who also have innovative solutions in mind to address these problems are the most valuable. Often these individuals have been commiserating with their peers about ways they might get their jobs done more efficiently. They are often the essence of a successful program save! (See Figure 5.2.)

With few exceptions, these special team members must be part of the program leadership team during the execution of a new Program Plan. They have a high level of product understanding and the vision and focused passion to

Enterprise Sponsored Leader

- May have past enterprise training.
- May have leadership experience.
- Demonstrated acumen.

Most Knowledgeable Leader

- Excellent knowledge of product and immediate task.
- Highly respected by team.
- Example to team of value of achieving work promised.
- May not be an obvious candidate.
- Assignment may germinate a highly capable future leader.

Figure 5.2 Look beyond the traditional sources for task leadership. The most knowledgeable can often derive the shortest path to success.

succeed. They are usually a major part of charting the most expeditious path to program success.

This source of highly valued leaders can be from a wide range of ages, personalities, and time on the program. They may not appear to have the typical acumen for a management role; however, it's clear that they have the knowledge and aptitude to drill deeper than anyone else to get the work in their subject area done quickly and completely.

It is essential that these special professionals be placed in leadership roles because of their expertise. This will be in effect a field promotion!

Some of these special individuals may at first be surprised and even uncomfortable that they are being asked to be a leader. They may not have been considered by the original program team or enterprise executives as an obvious candidate for leadership. They may never have envisioned themselves in this role. They may be a little overwhelmed by the commitment required to succeed in this new assignment. But they are most often the best person for the job of helping to save the program. If selected with objective, results-oriented criteria in mind, they and the program team will agree with the choice and support it. I have consistently observed that new leaders promoted this way promptly acclimate to their new role and do a very good job.

Sometimes, this highly knowledgeable candidate for a leadership assignment may exhibit some characteristic that disqualifies them as a lead. They may have a difficult time communicating work status or new ideas. They may have a personality that is offensive to some people, particularly when they are under pressure. But I have found these grounds for disqualification seldom occur.

There are a number of important side benefits to the enterprise resulting from looking for leaders from the subject matter experts on the Program. These efforts can identify excellent candidates for future leadership roles—some of whom would not have been obvious otherwise. This can be a good way of finding "uncut diamonds" among the ranks of enterprise employees.

Also, selecting results-oriented professionals to lead turnaround work is a strong and visible message to the whole Program team. It reemphasizes the value assessed by enterprise leadership of driving the program work to closure.

5.5 Establish Presence

The way the new Program Plan is introduced to the Turnaround Team is very important to ensuring that the program promptly and completely embarks on the turnaround work.

The first step is to conduct an "all-hands" Turnaround kickoff meeting. The purpose of this critical meeting is to introduce the Turnaround Lead and the new Program Plan—the Turnaround Plan—to the program team. This meeting should be attended by enterprise senior management/executives, large and/or key subcontract representatives, the prior program manager, and all members of the new leadership team. A representative from the enterprise Human Relations group for large enterprises might be invited to further validate the certainty of the new Program Plan. The customer must be informed of this kickoff. But if the presence of a customer representative might cause some team member to be shy about asking questions, the Turnaround Lead and the customer should consider the benefits of not having the customer attend. In either case, the customer must be provided with the meeting presentation package and meeting minutes, including actions promised, and they must have it briefed to them if they wish.

The importance of having visible and unquestionable support of the Turnaround by all enterprise management, enterprise senior professionals, all subcontractors, all suppliers, and the customer during the start of the Turnaround cannot be overstated. This is essential, even if the planning details have not been completely agreed to by all parties. It is most important to establish the new momentum of performing the new Program Plan immediately. Refinements to the plan can be made later.

As mentioned earlier, the new Program Plan and work tempo will be an abrupt change, to many members of the original program team. They will naturally want to make sure the new direction is "real" before they make a major change in the way they support the Program. The potentially large magnitude of the changes to the original program may be uncomfortable for some team members and even unprecedented in the enterprise.

The execution of the new Program Plan must be started right after the Turnaround kickoff meeting. The Turnaround leadership at this point must be totally focused on accomplishing the newly planned work needed to achieve the Turnaround Commitment on time. By demonstrating this focus, leadership will set the example that achieving the milestones in the Turnaround Plan must be the first priority for everyone on the program team.

It is very important that at the start of the program save a precedence is set that work progress is not allowed to linger, even if the program is waiting for external input. For example, even if the teams are waiting for material or new team members to arrive, there is likely still work that must be documented or personal training that can be completed while waiting. The program momentum when starting to execute the new Program Plan must not be allowed to skip a beat!

The Turnaround Lead must be physically visible and accessible to the whole program team, not just during the start of the new Program Plan, but throughout its entire life. The Turnaround Lead must make it clear to the leadership team that in the future he or she will take license to attend any meeting or event supporting the new Program Plan.

When the new Turnaround Lead is seen on the program premises or in a program meeting, there should be no question who the new program boss is. Sometimes the Turnaround Lead will attend program events with the past manager of the program. If this past manager has the best outcome for the program at heart, this will not be an issue. However, if there is any visible friction or competition for authority emerging between the new Turnaround Lead and the past program manager, the past manager must be asked to refrain from direct contact with the program team members. Even a little ambiguity about who is leading the Turnaround will destroy the team members' commitment to saving the program.

Many times while meetings are being conducted by the different program teams as the Turnaround Plan is underway, the Turnaround Lead should be seen "making the rounds." The Lead may show up for a short time at any meeting and simply stand in the back of the conference room while observing the meeting intently. As a result, part of the program team will have seen the Lead in their meeting and will observe that this person both understands and is interested in their particular issues. In return, the Turnaround Lead will get a sense of the vitality and success of the meetings they attend, get to know the team individuals better, and learn firsthand about some program issues.

5.6 Establish the New Tempo

The change in work tempo required to execute the Program Plan must have been clearly described during the kickoff meeting. Often more work hours will be required from all employees than were needed before starting the Turnaround.

For the new Program Plan, the tasks accomplished must be measured by percent task completion or equivalent, not just hours worked. The series of work steps required to achieve each major schedule milestone can be displayed with the traditional GANT or PERT types of formats (see Figure 5.3).

The PERT format clearly shows the dependencies required for each task and the critical path (the series of interdependent task steps with the longest duration in the schedule flow). It is a more pictorial way of showing the program schedules.

However, the GANT format makes it easier to identify tasks that must be accomplished in parallel with tasks on the critical path, including those that may be falling behind. Allowing these parallel tasks to fall behind will hamper progress on future critical paths when they are needed.

Completion of work can be measured by the completion of the tasks scheduled at a more decomposed level. A good program will decompose the main

PERT tracking

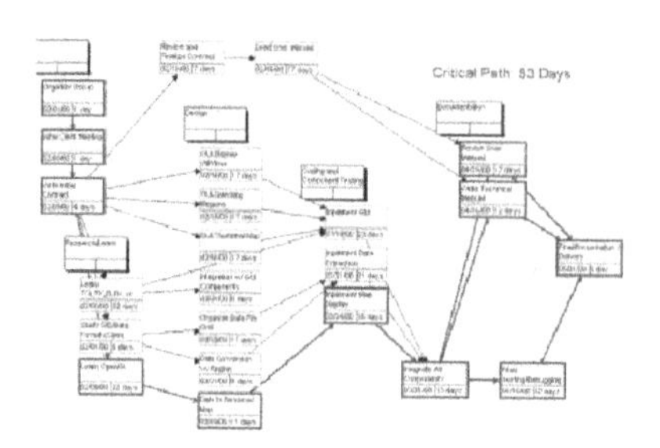

- **Clearly shows task interdependencies.**
- **Scheduling tools automatically identify the critical path.**
- **May be misused by just working the critical path.**

GANT tracking

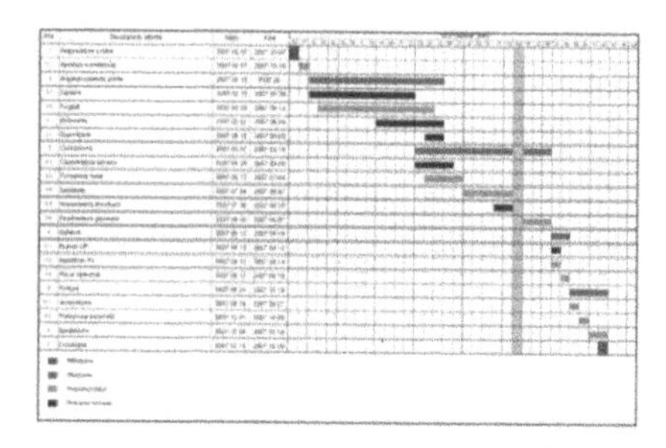

- **Clearly shows tasks planned to execute simultaneously.**
- **Easier to identify task work that may become the critical path.**
- **Helps leadership identify tasks falling behind that are not on the critical path.**
- **Better tool for "watching the flank" for yet undetected issues.**

Figure 5.3 Understand the strengths and weaknesses of traditional PERT and GANT program tracking formats. Each has specific values.

task to a level of subtasks such that one or more of these subtasks are scheduled to be completed each week.

Task work should be scheduled with sufficient margin to prevent anomalies such as late receipt of necessary input, work error, root cause/resolution, etc. from delaying the completion of the Turnaround Commitment. If schedule commitments are established assuming first-time success only, they will usually slip and become worthless as a tool to guide and motivate task closure.

During the Turnaround, the lowest decomposition of each schedule should be tracked daily at a program-level meeting, usually with the Turnaround Lead present. This will measure the work completed. When a team falls behind achieving a milestone (including the safety margin for unanticipated events), the team leadership must provide a recovery plan for getting back on schedule.

A schedule slip by one of the program teams must only be accepted as a very last resort by the Turnaround Lead, even if this task is not on the critical path. Slipping a schedule should be a very painful event for any program team and should seldom be accepted without the team first trying to successfully work around it. The program team leaders and team members must be prepared to create and implement innovative workarounds, work extra hours and make any necessary personal sacrifices to meet their schedule commitments.

The Turnaround Lead should consider using a schedule tracking system that monitors not just completion milestones but also work cost. There are excellent methods, such as Earned Value tracking, which simultaneously tracks adherence to both schedule progress and program cost. Care must be taken to make sure that schedule-only and cost-only information can be extracted from these methods; otherwise, controlling program-schedule and program-cost progress may become difficult because of the ambiguous status of each.

The Turnaround Lead and the program leadership team must be outstanding examples of making sacrifices and putting in the time to achieve their schedule commitments. Their behavior must underscore the critical importance of meeting promised completion dates. They must strive to never accept schedule slips for any deliveries they are personally providing. They must demonstrate creative workarounds and any necessary special work efforts to deliver their work on time.

The attitude exuded by the Turnaround Lead is always important, especially at the start of executing the new Program Plan. As mentioned earlier, this example establishes a precedent that the program team will observe and follow.

The Turnaround Lead must never let the concept of program failure enter their conversation. The Turnaround Lead must imply in all their interactions with the team that failure is never an outcome that will be considered. The Lead must display conviction every moment that program success is inevitable, although it will likely take a lot of hard work. The Lead must always be positive, supportive, and focused on the details to achieve the Turnaround Commitment.

As the execution of the new Program Plan gets underway, the Turnaround Lead must refer to just the one responsible individual for the status of each program task. If leading a team, this person's name should be on the organization chart for the Turnaround. Every team member should know who is ultimately responsible for the success of each program team. These actions will reinforce the policy of having one responsible individual for each program task (see Figure 5.4).

Also, the Turnaround Lead should always emphasize the importance of innovations for continuously improving work quality, simplifying work plans, and shortening work schedules. The program team members will think more about innovating improved process when they observe that this kind of thinking is recognized and valued.

The team will see that most of the time a proposed innovation, improved method or process does not result in the anticipated improvements or savings to the program. As a result, these proposed changes will not be incorporated into the Program's future. But the team will also observe that the benefits from

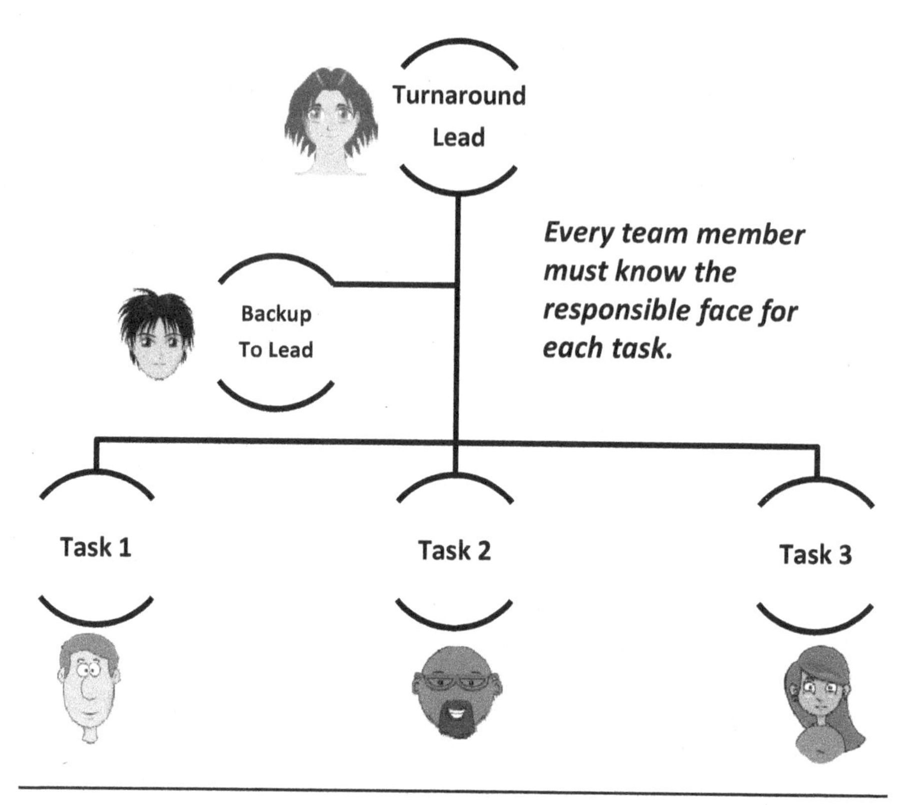

Figure 5.4 There must be one lead person for each task—not a team. When accountability is questioned in an all-hands event, everyone should point to the same person.

the innovations that *do* bring value to the Program far exceed the cost and time invested in attempting innovations that were not adopted. This comparison is important to convince the team members to keep looking for ways to improve program development speed, efficient use of cost and time, and product quality. Being on the lookout for these improvement opportunities should become a part of their daily work.

As will be discussed in more depth later, a program team begins to become on step when they start to appreciate the successes and accomplishments they achieve per the new Program Plan. Therefore, it is imperative that the Turnaround Lead share the team's accomplishments as soon as they occur. These compliments can be provided via an all-hands meeting, emails, special celebration events, or eye-catching posted notices in the work area as examples. The difficulty of the accomplishment should be stated, its value to the Program highlighted, and credit given to the team members working together to achieve the accomplishment. The feeling should be left with the team that they are special and that, with their high level of expertise, their serious work ethic, and their focused team cooperation, they can accomplish almost anything!

5.7 Easiest When Relationships with Team Individuals Are Not Yet Established

As discussed earlier in this book, the Turnaround Lead and new members on the program leadership team will find it easiest to make changes in staff assignments and program organization early in the Program Turnaround effort. This is because at this stage the new leadership will not yet have developed personal relationships with individual team members. This makes it easier for the new leadership to make objective choices for the best individual for each assignment based on documented accomplishments. It also allows the leadership to make assignments without having to answer to any developed friendships.

Furthermore, at this time there will be no basis for a team member to assert that an assignment was made by a program leader based on favoritism.

Program leadership will always have to navigate a narrow path between being cordial and friendly with all team members while not paying special attention to individuals they privately feel could be good friends.

5.8 Share the New Organization with the Team

Changes of work assignments and reorganization of the Program must be implemented promptly so executing the new Program Plan can start immediately. Any delays permitted by the program leadership to allow the program team members to comment on or challenge the new organization and/or assignments can damage the initial fast progress needed to stay on track with the Program Plan. This can even cause the Program Turnaround to fail.

Almost every program team member will have a different view of and preference on how the new program turnaround is organized. No organizational structure will make everyone happy. The new organization must help directly address the "cavities" found in order to achieve the Turnaround Commitment. The newly established program design must start its work with the organization and assignments provided by the revised program turnaround leadership. All Turnaround Team members should be reminded the initial program setup may not be perfect. Corrections and improvements can and will be made later.

I highly recommend that the new program leadership not share the new organization with the program teams until after it is approved by senior enterprise management, the customer, and the Program leadership team—plus possibly some of the major subcontractors at the discretion of the Turnaround Lead—and placed under configuration control.

Some team members from the original program may be disoriented by the rapid change of organization. This is normal, and these team members will usually acclimate to the changes within weeks. As discussed in the following section, it is better for the team to endure the discomfort of change all at once.

Changes to program leadership assignments should be shared with the Program leadership team before the new organization is announced. But the program leadership must keep this knowledge to themselves until after the new organization and leadership assignments are presented at the all-hands kickoff for the new Program Plan.

5.9 Take the Medicine All at Once

It is important to make the transition to the new Program Plan all at once. Allowing critique of or pondering over the changes by individual team members will only confuse and delay critically needed progress on the new schedule to save the program. Program leadership must reassure the team that critique and recommendations will be heard and incorporated at the appropriate time.

Transitioning to the Turnaround Plan all at once will lessen the total anxiety and uncertainty experienced by the team members. It is possible that some team members will continue to question the changes, and there may even be a temporary setback in team morale. Making all the changes at once will minimize this impact to the Turnaround schedule. These savings derived from making the program changes at one time similarly apply to the work performed by the customer, enterprise executive management, other enterprise programs, subcontractors, and suppliers.

5.10 Maintain Seamless Momentum and Focus—"This Is Serious"

It is important that all members of the program team make the transition to the new Program Plan together. There must be no differences in the amount subscription to the new plan as a function of seniority, time on the original program, or other reasons.

There must be relentless reminders by all program leadership that the Program has embarked on a dramatic change to save it. There is no time for a gradual transition period. Plus, a gradual transition will cause expensive confusion. This leadership guidance must be reinforced by leaderships example of abiding to the new activity level, determined by the new Program Plan. Everyone on the Program must know that the Program Turnaround is real, serious, and underway!

5.11 Chapter Highlights

- Turnaround Leadership must evaluate the program quickly.
  - Review program promises.
  - Identify the cavities.
  - Identify the major changes needed
- Immediately determine draft organization.
  - Review, correct leadership and assignment deficiencies.
  - Review, correct headcount and talent mix deficiencies.
  - Update reporting structure if needed.
- Make field promotions.
  - Task leadership ability often trumps management experience.
  - Best leader might not be a candidate from the enterprise manager development track.
  - Might identify hidden high-capability leaders.
  - Emphasizes the importance of results to aspiring leaders.
- Establish presence.
  - Be frequently visible to everyone on the Turnaround Team.
  - Show solidarity between Turnaround, customer, and enterprise management.
  - Start executing the new Program Plan immediately.
- Set new Program tempo.
  - Measure work by tasks completed, not hours worked.
  - Establish a positive attitude that does not question success.
  - One person is responsible for each task.
  - Innovations and continuous improvements have high priority.
  - Promptly celebrate even small team successes.
  - Leadership must set a constant example.
- Report new organization to the program teams as soon as complete.
  - Supports prompt start of Program Plan.
  - There is little time for rebuttal; organization can be refined later.
- Take medicine (start new plan) all at one time.
  - Reduces costs and confusion caused by adjustment to new plan.
  - Savings also apply to the customer, enterprise, subcontractors, and suppliers.

- Settle down to new place quickly.
  - Program leadership must be an example.
  - Shortens time to achieve the Turnaround Commitment.
- Establish new momentum and focus in all team members.
  - This is serious!
  - All leadership must clearly support the new plan and tempo.

Chapter Six

It's a Campaign, Not a Program!

The members of the team saving the program must be encouraged to have pride and awareness of the high value of their accomplished work. This is usually well deserved because of the extra sacrifices the individuals on the team are making. Leadership must consistently acknowledge that the Turnaround effort is a special activity and therefore the program team is special.

Team members must be recognized for much more than just working arduous additional hours. They also must be appreciated for having a high degree of understanding about the problems they are solving, for having a highly competent level of control for what they are doing, for developing highly creative solutions to problems, and for achieving important milestones on time. Team spirit must be rallied and brought to an extraordinarily high level to muster the extra energy needed to get the troubled program back on track.

6.1 First Time in History

A program team that is saving a development program must be reminded that they are very likely the first in history to be developing their product or refining it the way they are. In addition, because what they are developing is original, the team may need to develop new methods and processes to accomplish their work. They should be reminded that their work requires exceptional creativity and effort to get the program back on track (see Figure 6.1).

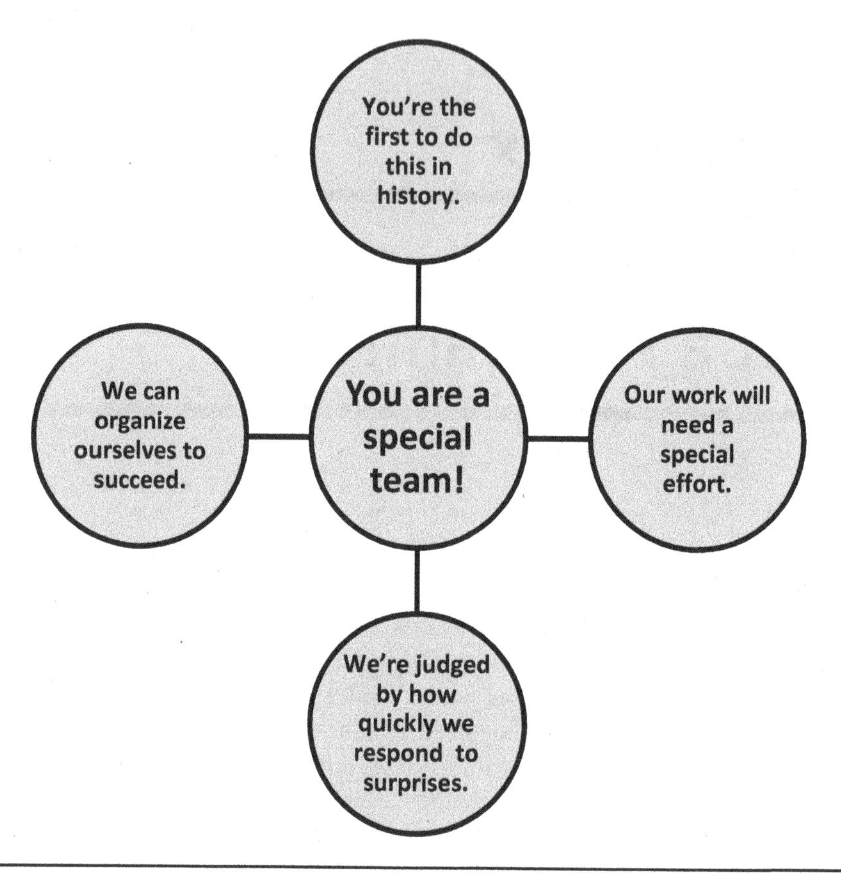

Figure 6.1 The Turnaround Program team must be reminded that they are making a special contribution, requiring above-average team members. Extra hours and personnel sacrifices will probably be necessary.

The team must be reminded that their performance will be judged not only by how accurately they execute the new Program Plan, but also by their ability to responding quickly and effectively to development anomalies and failures. These will occur. The special importance of correcting Program test and other anomalies and failures on the first try, and providing a method for doing so, will be discussed later.

The program team should also be given latitude to organize themselves for doing their work and managing their problem solutions. The Turnaround Lead and team leadership have selected team members with the required talent to get the program back on track. The wise leader knows that often a high-caliber team, focused on results and task closure, will organize themselves better than the team leaders can. The experienced lead/manager must judge to what degree to keep their hands on the reins when allowing the team to self-organize. But giving the teams some freedom to do so will enrich team morale and improve program performance. The amount of freedom allowed will depend on the maturity of the team. A lot of excellent junior team members like the opportunity to show the boss they can plan it better!

6.2 It's a Lifeboat!

Imagine you're on vacation when your cruise ship hits a reef and quickly sinks! Fortunately, all hands got onto a lifeboat that now is drifting in the sea. Your first thought is, "Thank goodness I got into this boat before the ship disappeared beneath the waves. We're safe!"

But then you notice water sloshing around in the bottom of your lifeboat. You also see some bailing buckets hanging from under the seats. The lifeboat is bobbing around in the heavy seas, and more water will likely be entering the boat.

So what do you do? Call a general meeting of the occupants to perform a trade analysis of what options could be pursued? Remind the team you are more senior than most of the other occupants of the boat and removal of the water should be performed by the freshman occupants? Move yourself to the bow of the boat so you may identify yourself as one of the senior leaders of the boat? No! Your grab the bucket and bail! Format is not relevant in this urgent situation—content is. This is an emergency!

This is exactly the attitude all members of the program team must have while executing the Turnaround Plan. This program is in a dire situation—that's why the remedial action you are a part of was necessary. It may be on the verge of failure. The program team members must all support each other in any way they can to keep the program from failing. If a team member offers to help, you as a program leader must immediately find a way that they can. Leadership cannot refuse assistance from the program team because of convention or traditional habits (see Figure 6.2).

You must not question how much prestige may be associated with any role you take to save the program. Just do whatever it takes. The program must achieve certain promised milestones to achieve the Turnaround Commitment or it, and part of the enterprise, may be lost! Your enthusiasm and adaptability to perform the next urgent task as a team member or Program leader will impress and motivate the program team and members of the sponsoring enterprise to do the same. A complete focus on achieving the milestones necessary to save the program must be exemplified by all Turnaround Program leadership.

- **"We must work as a team to survive."**
- **"We must assume any role that is needed."**
- **"Our most important task is the one we are working on now."**
- **"No time to question the prestige of our role or workplace."**
- **"Sometimes our effort will not be perfect but it must be sufficient."**
- **"Meeting our commitments is necessary to survive."**

Figure 6.2 Turnaround Team members must think as if they were on a lifeboat. The tasks most critical to survival must be performed first. The team depends on each member working the most important thing needing attention *now*!

Keep in mind that sometimes the program solutions derived in this environment are not perfect. For example, there may end up being more safety margin than necessary in that new mechanical design, or it may not have been possible to find the targeted number of math teachers with the appropriate skills to design an improved math curriculum for the fifth graders. However, in almost all these cases the solution will be adequate. The Turnaround Team must quickly demonstrate that they have a way to revive the program and achieve its promises. Work beyond what is required to accomplish these adequate needs—if of future value—can wait.

6.3 Support from Executive Management

It is imperative that the enterprise executive management completely support the Program Plan and the work needed to accomplish it. Any reservations that may linger among the individual executives must be kept at just the executive level. The new plan will often require the program team to take unconventional, risky, and sometimes unprecedented actions, and they will need the full support of the enterprise.

The enterprise support should not just be for the program currently being saved but also for the basic concept of a special team making extraordinary sacrifices to save any enterprise program. Enterprise management should convey their appreciation of the extra dedication and effort put forth by the team members in this campaign. They must be forward in conferring special recognition for exceptional accomplishments by the team as they execute the new Program Plan. This recognition can be provided to both team individuals and the whole teams.

Some large enterprises are coy about highlighting a new Program Plan to save a program in their portfolio. This may be because of fear that doing so will advertise a failure in the enterprise. The news of serious problems in a program travels very fast in any industry and is difficult, if not impossible, to mediate. Enterprise leaders are often surprised by how fast the rest of the industry knows about their problems. However, for a business to demonstrate that they have promptly recognized that one of their programs is falling behind and has taken prompt, processed-based action to correct it can be very beneficial for them. It demonstrates their high dedication to providing what they promised to their customer, as well as their high level of executive vigilance.

6.4 Not a Maverick

The members of a high priority Turnaround effort must not be allowed to be arrogant. They are a special team, given special emphasis by the enterprise to get a troubled development program back on track. However, they must not think of themselves a group of "rebels" forging a new way to conduct business by the enterprise.

The team must be reminded that, because a new Program Plan and changes in the program organization were needed for a program in trouble, this action does not in any way discredit the other programs being executed in the enterprise. Nor does it lessen the importance of the issues the other programs are solving.

It will be helpful for the Turnaround Lead to occasionally inform the teams of some of the challenges the other programs in the enterprise are working on. These are challenges that are common in most programs and require creativity and hard work to solve. They typically include acquiring sufficient funding, receiving priority to access key facilities, assembling the correct talent mix for their work, and more.

6.5 Probing and Gossip

Many times a Turnaround Plan and program reorganization will be looked upon by outsiders as a panicked, last-ditch effort to save a program. They may feel that any success this new team has is just attributable to exceptional team members and intensive, special management supported by the enterprise. Many of these outside people may feel that they must defend the way their own program teams perform their work when this Turnaround Team demonstrates success.

This is a natural human reaction by the outside programs. Properly handled by enterprise leadership, it can actually be used to enhance the success of the program save and improve the other programs in the enterprise.

First, a quick review of who some of those might be that are commenting.

Sometimes senior employees who have been working for the enterprise a long time find the unconventional approaches in a new Program Plan to be disconcerting. They likely have experienced numerous successes in the enterprise using more traditional and tested program-planning approaches. Seeing a high-intensity, possibly quite innovative, Program Plan being attempted, usually with many new team faces, can add to their concern.

As discussed earlier, team members of other programs in the enterprise may view the new Program Plan as highly unconventional, even experimental. They recognize that their own program has succeeded following the more traditional path. In addition, they may question why a program that has caused so much trouble merits getting so much attention and support from enterprise management. They may question how the new Program Plan can be in any way superior to the program plans for the other enterprise programs. After all, the other program plans were derived from the enterprise's proven and highly refined processes.

Sometimes members of the enterprise executive team, many of whom were once program managers, will question the approach of the attempt to save the troubled program. These executives know they are being observed by employees of the enterprise for their reactions to these attempts. The keen observer may detect some caution and reserve in their support of the Turnaround Plan. These executives understandably do not want to leave the impression that they are abandoning any existing enterprise processes without very thorough scrutiny.

Consideration should be given to the reactions of the major program subcontractors and suppliers to the steps to save the troubled program. This is especially true for those subcontractors who have successfully completed work for other programs in the enterprise. Sometimes these subcontractors and suppliers have established a precedent of helping to influence the direction of the enterprise programs they have supported. There may be established personal relationships between managers in their subcontractor organization and the enterprise. The different organizational structure and high work pace associated with the Turnaround Program may challenge the influence they have had in the past.

This can lead to probing and critical comments from subcontractors and suppliers. To protect their past relationships with the enterprise, they may show empathy toward those in the enterprise who are resisting the new Program Plan. They may assume the role of the "realistic" participant in the Turnaround planning. They may portray themselves as seasoned overseers of the Turnaround effort, not allowing themselves to be swept away by the high energy and enthusiasm it has engendered.

These kinds of responses from subcontractors and suppliers can be dangerous to the success of the program save. In the extreme, they can result in the business interests of the subcontractor or supplier superseding the success of the Turnaround Program. Correctly structuring the teams and interfaces that manage the subcontracts and suppliers will help ensure that these supporting organizations are devoting all their efforts toward making the Program Plan successful. This will be discussed in more detail later in this book.

On occasion even the customer can be a source of probing and criticism regarding the program save, usually not from the employees directly supporting the program, but from other employees in the customer's organization. It is human nature for the customer to question the attempt by its contractor to save their program. Some customer personnel may be concerned that the contracting enterprise is making a panicked attempt to save a program that some might believe will inevitably fail. They may be afraid that this attempt will end up wasting more time and money. It must be the responsibility of the customer's leadership to disarm any Program concerns or criticism from within their own organization.

6.6 Schedules versus Diplomacy

Of course, probing and gossip about the new Program Plan and the attempt to save the program will naturally persist from sources outside of the Program. It is human nature to comment on what may appear to some to be a highly energetic and even unconventional reaction to a problem. However, some actions can be taken to reduce and even take advantage of this scrutiny. This bit of diplomacy may require a small investment in time by the Program but may reduce the overall burden of achieving the Turnaround Commitment.

First, the program should, within the limits of protecting sensitive program data, educate the probing organizations about the approach and expected outcomes of the new Program Plan. The Program should then keep any sceptics apprised of the progress of the save. The Turnaround Lead should consider inviting these people to the internal status reviews, provide them copies of the program activity reports, and even provide in-person progress reviews within the limits imposed by any sensitive data. The purpose of this would be to provide facts, dispel rumors, and set the stage for asking for constructive critique and recommended improvements from the critics.

When providing progress summaries to the probing organizations, the Turnaround Program should summarize the lessons they have learned. The information should be presented with the desire to help keep the probing organization from making the same mistakes. This can improve the success of the other enterprise programs and projects.

When feasible, the program should offer tours of their program work area(s), facilities, and implementation sites. Again, while doing this any sensitive data must be limited to only those who have authority and need to access to it.

Always solicit suggestions for improvements from probing organizations. You may in fact help their efforts by sharing your approaches to the program save and by sharing the lessons you have learned from the Turnaround Planning. They may help improve the new Program Plan with their critique and suggestions. This exchange can turn the natural probing and commenting from other organizations into a mutually beneficial relationship.

6.7 Chapter Highlights

- "We're the first ones in history to do this!"
  - Judged by our ability to respond to issues quickly and accurately.
  - Some special freedom to self-organize.
- "We're in a lifeboat!"
  - May have been on the verge of failure.
  - Must assume any role that is needed!
  - Sometimes "good enough" is sufficient.
- Support from enterprise executive management is necessary.
  - Provide special recognition.
  - Show no hint of skepticism.
- Expect probing, gossip, criticism.
- Trade effort versus benefits to educate those not on program.
  - Share lessons learned.
  - Tours.
  - Solicit valuable critique.

Chapter Seven

High-Value Elements

This is one of the most important chapters in this book. It identifies key elements needed to make the Turnaround successful. Many of these may be features beyond the scope of what is specified in the Program Plan. They have been found to be effective from years of experience successfully managing many different programs and program saves by many different program and project managers. They facilitate strong forward progress during execution of the Turnaround Plan (new Program Plan) to save a program and minimize the need to redo what has already been successfully completed.

7.1 Face to Face

By far the most effective way to communicate between two people is face to face. This is the irrefutable result of hundreds of thousands of years of evolution of our species.

Science repeatedly tells us that 70% to 80% of face-to-face communication is nonverbal. Any good salesperson, politician, or program manager knows this. Scientists are still scratching their heads about why so much more is communicated face to face. Are all five senses somehow involved? Is there somehow an added dimension of trust established with someone who shares space with you? For now, we should accept that it is a fact of our human nature, and capitalize on it for the fast and efficient communication we need during a Program Turnaround (see Figure 7.1).

- **70% to 80% is nonverbal.**
- **Provides immediate access to current information and responses.**
- **No delays present, as with emails and notes.**
- **Uses up to four more human senses than a phone call.**
- **Builds the highest level of trust between team members.**
- **Easy to use by a collocated team.**
- **Applies a highly effective communication system that has evolved over 100,000 years!**

Figure 7.1 Face-to-face communication is quickest and most complete. It's how we humans evolved. A good leader capitalizes on this for their program or project to be on step.

"Skunk Works"-type programs* have benefited from this human characteristic for over 70 years. They facilitate face-to-face communication by the colocation of program personnel. This will be discussed in more depth.

The successful Turnaround Program will facilitate face-to-face communication among all team members to the greatest extent possible. This type of communication is imperative for any program to get and stay on step. Handwritten notes, email, texting, and other forms of written correspondence may be necessary if, for example, complicated data must be conveyed or some form of formal obligation must be recorded. Otherwise, being present with the person you want to work with and communicating face to face is by far the most complete, accurate, and fastest way of conducting business.

When discussing something face to face, questions can be asked, responses are immediate, and the people involved will naturally concentrate exclusively on

* Skunk Works is a highly efficient development approach that has delivered many exceptional aircraft to the U.S. Government. They are noted for providing aircraft with outstanding performance using small teams and short schedule durations.

the subject at hand. Responses to questions, the acceptance of proposed ideas, and more will be evident much more quickly and with less ambiguity than by just a written or verbal comment—a high degree of trust and commitment is established between the people communicating.

One example of the terrible tragedies of business via social media is what I call "the five-day email." I'm afraid we have all encountered this misuse. Often it starts by your being cc'd on a lengthy email question-and-answer dialog by two individuals who may in fact be a short walk from each other. It might go something like this:

"Hey Jim, before I order part 749 I have to know if you want the –A version or –B version."

4 hours later,

"Hi Bob, I had no idea there was a –A or –B suffix for this part. What do they mean?"

Following day,

"Hi Jim, –A is plastic and –B is Teflon."

6 hours later,

"Bob, still don't get it. What is this material for? Insulation?"

Next morning,

"Hi Jim, I'm sorry. This is the material to cover the terminal ending."

Noontime,

"Hey Bob, why would anyone use Teflon for a terminal ending? Should I be interested?"

And on and on and on it goes!

Jim and Bob may believe they're using an acceptable communication process, that they're incorporating modern technology. They may believe they have saved time and cost for the enterprise by not walking downstairs or to the next building to talk to the other person. Instead, they just type a note, run spell check, and hit "send."

However, I and everyone cc'd wasted valuable time reading an email dialog that spans days and has little if any value to the rest of us. Even more important, this dialog could have been concluded with a short walk and a three-minute face-to-face conversation. The underlying issue could have been completely resolved in less than fifteen minutes instead of several days, and Bob and Jim would not have wasted the valuable time of other program team members. If the resolution of even small issues is not resolved as quickly as possible for a development program, it will be impossible to achieve the major program milestones quickly.

There are other instances in which attempting to incorporate modern computer-based tools turns out to be a misuse of them—for example, conducting virtual computer-based meetings among program team members with offices in the same building and even on the same floor. Or catching up on emails while attending a program meeting. Or team members attending simultaneous meetings via computer and multiline phones from their office and calling it "multitasking." Or remotely performing analysis and planning via a common program database while some participants don't understand the purpose of this work or how to use the database tool.

The maximum use of high-speed, low-cost face-to-face communication is absolutely necessary for the time- and cost-critical environment of a program turnaround. In addition, a face-to-face transaction increases trust between the team members, which improves all team performances. It is necessary for a project or program to be on step.

7.2 Virtual Communication with Care

I discussed above the advantages of face-to-face communication when executing any plan for a development program. This is the fastest way to build trust and solve problems in a team. Non–face-to-face communication such as handwritten notes, emails, texts, phone calls, etc. can be a disadvantage during a program save.

But how about video teleconferencing (VTC)? With increased camera pixel density and higher resolution displays, are we approaching the point where the flat screen will provide the same information as face-to-face communication? After all, it could save time and money not having to walk distances to meet people on other parts of the campus, not to mention the huge amount of time and money that would be saved avoiding traveling to somewhere. For instance, if you live in the US and travel to the opposite coast for face-to-face discussions, it will take two to three days, including travel, to conduct even a one-hour meeting.

Unfortunately, tests show that high-definition VTC does not make up for the absence of physical presence. So far, there appears to be no way to acquire that 70% to 80% nonverbal communication premium without a face-to-face in-person meeting.

This does not mean that all communication for a development program on step must be face to face. When separate activities on a program are "loosely coupled" (when two portions of a program exchange little data and have little or no interdependency), information can be exchanged by non–face-to-face means with little detriment to program schedule. A face-to-face introduction of the respective leads for these program activities at an opportune time, such as the kickoff of the new Program Plan, may by adequate to execute this loosely coupled interface successfully. This one face-to-face meeting will develop a foundation of trust between these leads.

7.3 Colocation

The importance of collocating the members of the Turnaround Team cannot be over-emphasized. By collocation, I mean all that team members work in the same building enclosure and a short walking distance from each other. For large teams of hundreds of people, this may sound unreasonable, but it is not. Very often the enterprise can fill a very large work area with many cubicle-type partitions but without any complete floor-to-ceiling walls dividing the various program teams, the cubicles providing the needed privacy. An acceptable noise level for working is maintained by professional courtesy. Yet solutions of development challenges via direct contact between members are immediately available (see Figure 7.2).

This environment may not be conducive to leaders who believe they must assert their value to the enterprise by the elegance of an elaborate private office. But it is the way successful development programs and turnaround programs have gotten on step and stayed there for many years.

All the specialists on the core project or program team needed to achieve the Turnaround Commitment should be sitting in this common area. For large programs this can include Human Resources personnel, the business team, quality engineers, design engineers, test team, management, manufacturing planning personnel, machinists, software coders, and so on—all the functions and expertise needed to achieve the committed outcomes.

Colocation provides access to face-to-face communication with the team members responsible for any of the program functions, by any program team member needing to communicate with them. Team members can initiate this communication and solve problems as quickly as taking a short walk.

The fact that seating assignments are more homogenously distributed in this collocated work area deemphasizes the distinction of organization structure and employee seniority. This encourages the team members to approach each task more as a group of equals.

- **Facilitates easy face-to-face communication.**
- **Facilitates fastest solution of development problems.**
- **De-emphasizes individual distinction and organization structure.**
- **All functions necessary to achieve Turnaround Commitment should be present.**
- **Facility can be "basic," as long as it's safe, clean, and comfortable.**
- **Remember the "30-second rule."**

Figure 7.2 Colocation facilitates face-to-face communication with all team members. It provides immediate access to team members and reduces the cost to integrate work products.

This use of colocation has been successfully applied by startup and product development teams of all sizes in Silicon Valley and the US as well as large and complex development programs at Lockheed Martin and similar high-technology organizations, where it has been shown to be essential to turning around a troubled program and getting it on step.

There will be times when the needs of the customer, such as the need to distribute program work as in the case of government contracts, require that the program work be geographically distributed. If the customer for the Turnaround requires that portions of the Program be developed at separate facilities, a high degree of effort must be devoted to assuring that the tasks performed at these places are as independent of each other as possible. This includes choosing divisions that require that the separated portions share few and infrequently communicated work products. This need is discussed more later.

7.3.1 The 30-Second Rule

As a program manager, I often established a "30-second rule." I would stipulate that I should be able to briskly walk to any team member on the main program site and start a face-to-face conversation in 30 seconds or less. Since I was usually seated to one side of the program work area, any member of the program could reach me or another member of the team within 30 seconds. Planning work and resolving issues were available quickly. If multiple specialists were needed to solve a problem, they could all be pulled together immediately. Good ideas could be quickly addressed in an informal environment. All communication was face to face, providing short times to closure. A high degree of team trust was established. This made it much easier to get the program on step and stay there.

I vividly recall a special example of how valuable colocation is. On this program, we had a simulation group who needed access to their special lab facilities, which were located a little over a one-minute walk away from the rest of the program. This lab was separated from the rest of the program by multiple floor-to-ceiling walls. The team working with this equipment chose to situate their desk area next to this equipment, as it made perfect sense to have their computers, phones, and desks close to the equipment they were using.

Yet at the morning program status meetings, this simulation team seemed to be out of touch with the rest of the program. Their leaders sometimes showed up late, and the team was often unaware of the current events in the program. They did not attend other program reviews and functions they could have benefited from. This did not allow the development program they were supporting to proceed at the high rate it was capable of.

At first I thought the simulation team had a history of being independent before I arrived as the new Program Lead. Maybe this was why these team members were not quite as closely in step with program activities as the other teams. But then I realized it was simply because they were not collocated with the rest of the program.

So I asked the simulation team to move their desks and sitting area to be physically collocated with the rest of the program team. This meant now they would have to walk one minute to their simulation lab in a different room to work with the lab apparatus.

This simulation team was understandably appalled. They thought the Program Manager might be off on some idealistic tangent! But within a week of the move, a change occurred. The scheduled activity and issues being addressed by the simulation team were now in lock step with the rest of the program. The simulation team lead and assistant attended all the needed daily and weekly program meetings. The simulation team members were now aware of critical program meetings and other team events and participated in them. Most important, the simulation team members now contributed much more to program problem solving and planning. As a result, the program started achieving its commitments at a faster rate.

Colocation is a critical consideration for a Turnaround Plan that cannot be overlooked. It facilitates our fastest and most effective human team traits.

7.3.2 Minimal Information Coupling and Maximum Functional Cohesion When Separating Program Teams

There will be times when the needs of the customer require that the program work be geographically distributed. For example, it is completely understandable that if an important US government-funded development program is looking for congressional support, they spread the benefits of the work to as many states and congressional districts as possible. This is good for the U.S. economy. Unfortunately, this goal can work against the advantages of collocating the work.

If it is necessary for portions of the Turnaround Program to be geographically separated, any negative effects caused by the loss of colocation can be lessened by how the portions of the program are divided among the different work sites. The loss can be reduced if the pieces of the work assigned to the different sites are chosen so they are as independent of each other as possible (see Figure 7.3).

The measure of the degree of independence of the functions in a system is becoming more quantifiable. Starting in the 1960s, methods were developed to

POORLY PARTITIONED WORK

Manufacture and Test New Drug
Miami Florida
Manufacture and Test West Coast
Los Angeles California
Manufacture and Test East Coast
Norfolk Virginia
Evaluate Test Results
Seattle Washington

- **Manufacture and test have loose relationship.**
- **Much data between three subordinates to coordinate and complete.**
- **Prone to errors.**

BETTER PARTITIONED WORK

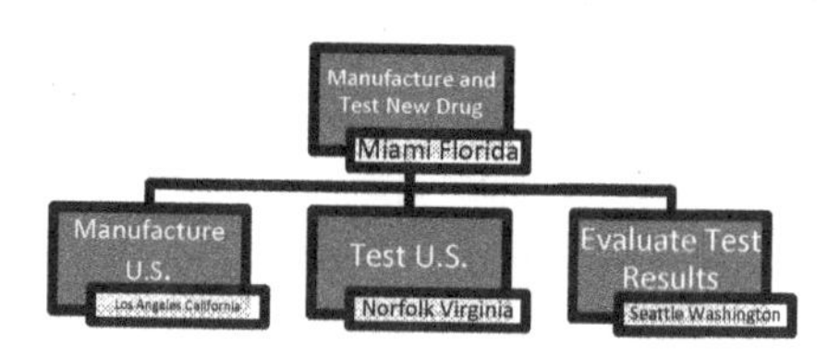

- **Functions in each subordinate are highly cohesive.**
- **Exchange of only conclusive and complete data and products between subordinates.**
- **Less error prone.**

Figure 7.3 Teams and facilities that must be physically separated should perform work that is as independent from the work of the other sites as possible. A comparison of maximum functional cohesion in each site and minimal interdependencies (coupling) between sites can be a useful tool for trading the candidates for dividing the tasks between each site.

best

LEVEL	DEFINITION
Functional	All parts contribute to a single well-defined task.
Sequential	Parts are together because output of one part is the input of another.
Communication/ Information	All parts operate on the same data.
Procedural	All parts follow a certain sequence of execution to complete a task.
Temporal	Parts are only executed in a fixed sequence in time.
Logical	Parts are logically categorized to do the same thing even though they are different in nature.
Coincidental	Parts are grouped arbitrarily.

worst

Figure 7.4 Trying to achieve maximum functional cohesion in each part is a good guide for dividing something into its most independent pieces. It works with trying to achieve minimum data coupling between the parts. This hierarchy was originally created to assist software design and will be discussed more in Chapter 15.

divide a complex social organization or large software program into pieces that were as independent of each other as possible. One such method, developed by Edward Yourdon and Larry Constantine,* has the user grade the functional cohesion of a subportion of a system (how alike and interdependent the functions in the subportions are) and the degree of information coupling among these subportions (the amount of data conveyed in this coupling and the extent of control information it contains). The approach is that the more you can partition these subportions so that functional cohesion is high and interface coupling is low, the more independent these system subportions will be.

Please see Figures 7.4 and 7.5 for the definitions and hierarchical levels of functional cohesion and data coupling. When necessary to divide the work of the Turnaround Program to be performed at geographically separated worksites it is essential that the work at each site be as independent from the others

---

* Yourdon, E. and Constantine, L. (1979). *Structured Design: Fundamentals of a Discipline of Computer Programming and Systems Design.* Prentice Hall.

best

LEVEL	DEFINITION
No Coupling	Modules do not communicate with each other.
Message	Modules exchange discrete state information in messages.
Data	Modules exchange data via direct communication.
Stamp	Modules use data in a common database.
Control	One module can control the functions in another.
External	Modules share an externally imposed data format, communication protocol, or device interface.
Common	Modules share the same global data.
Content	One module can change the internal workings of another module.

worst

Figure 7.5 Trying to achieve minimal data coupling between each part is a good guide for dividing something into its most independent pieces. It works with trying to achieve maximum function cohesion in the parts. This hierarchy was originally created to assist software design and will be discussed more in Chapter 15.

as possible. Establishing via analysis, the maximum cohesion of the work performed at each work site and minimum coupling of the data communicated between each site is necessary to get the program on step. This will help ensure that the staff and facilities resident at each site can focus on a set of highly similar and interdependent tasks, while being minimally distracted with managing data dependencies that need to be communicated among individual sites.

7.3.3 Organizations Mapped to Product Breakdown Tend to Run Most Efficiently

The approach of mapping the work of the new Program Plan to different geographical sites presented above also applies to the breakdown of the program into its departments and teams—the organization of the Turnaround Program.

There has been a debate in some businesses as to whether to organize a program around the functions needed to create the product or around the actual target elements assembled to create the product. For example, if the program is developing a new automobile, do you divide the program into an electrical engineering group, a mechanical engineering group, a stress analysis group, etc.? Or do you divide it into an engine group, a transmission group, a wheel brake group, etc., specific to the car?

The first of these is often called a *functional* organization, the second a *product* organization. For a large enterprise that builds many versions of the same product, a functional organization may be more efficient. For example, if your automobile company builds, say, 15 different cars, the economy of scale might make it most cost-effective to have a central mechanical engineering group that develops the mechanical designs for all the cars. Having one group doing this can capitalize on reusing design elements between the cars, assuring compliance to enterprise and government standards, establishing parts standardization, focusing the enhancement of new mechanical design tools and design methods in one organization, etc.

But if you are building your first and only one car for a new enterprise, there is not yet a large enough scale to capitalize on a functional organization. In addition, for a Turnaround Program the extra work communicating the need work to the functional organization, managing the interfaces of their work with the other functions required for the Turnaround Commitment, making sure the appropriate work priorities are established, making sure schedules are coordinated and more can hamper or even defeat achieving the Turnaround Commitment.

After all, when you were a child and you decided to pull together some of the neighborhood kinds to build a downhill racer, you wouldn't organize functionally. Similarly, regardless of whether a team is building the first personal computer, creating the first production jet fighter, or drafting the U.S. Constitution, a product organization is the *best* selection.

Often large enterprises with many development programs will use a hybrid organization of functional and product elements. In this case, the enterprise might keep a "central" organization containing groups of the various specialists needed for enterprise products, then these specialists are deployed to the various development programs. If the Turnaround Lead is running a program in this environment, there must be documented agreement with the functional organization as to how their personnel will be deployed. The Lead should stipulate that the Turnaround Program direct the work of the professionals loaned to them from the functional organization. The Program should insist that it have a major role in grading the work performance of these professionals and be guaranteed the use of their services for a specific length of time. The Turnaround Lead must share very little of the guidance and evaluation of their team members with any other organization.

Given that the Program Turnaround activity now will apply a product-based organization, how are the divisions determined? There are two ways that are immediately accessible.

First, the principles of *minimum coupling* and *maximum cohesion* used to distribute program work to different geographical sites may be used to divide the program work into product organizations. In the absence of other organizational direction (such as enterprise mandate, customer mandate, etc.), this is the best method for determining the program structure. It often starts by evaluating a range of proposed organizational structures, checking each design for cohesion within each department or team identified and coupling among the different departments or teams. Again, the hierarchies of functional cohesion and data coupling are shown in Figures 7.4 and 7.5.

The aim is to divide the Turnaround Program into pieces that are as independent of each other as possible. The result will be to greatly reduce cost and schedule time to reach the Turnaround Commitment by reducing the expensive coordination of events and data among program groups. The resulting program design also ensures that the team members in each program group are efficiently using their time to solve the same problem(s).

Keep in mind that much confusion and many mistakes will be avoided if simply the name given to each department or team in a Turnaround Program clearly describes what they do. For example, you can label a team "Dynamics" or "Documentation." But would it not make it much easier to remember what they do if you called them "Steering Stability System" or "Archival and Backup Documentation"? This higher degree of clarification will reduce the expensive mistake of team members' misunderstanding what team is developing what function. It will be surprising to observe how this clarification of team titles will even reduce confusion for people who have been on the team a long time.

A second source of guidance for dividing the program save into departments and teams is to mirror the way the customer has divided up the work when describing what work they want done. When a customer solicits and/or specifies a program or project to be accomplished, especially when this work will be managed via a contract, they will divide the work into what they consider to be its most basic and independent component pieces. This is sometimes called a *work breakdown structure* (WBS). Organizing the program to mimic or clearly map to the WBS of the customer will provide a valuable mechanism not only to assure that each specified requirement is being performed by the program, but also to know in which team it is being performed. This greatly reduces the chance of misinterpreting which team is completing each required task, a mistake which in past program designs has caused schedule delays, added program costs, and other detrimental mistakes.

Because of the customer's motivation to decompose the program work into pieces that are as easy to understand and as independent as possible, a WBS or similar customer document often provides a good program decomposition design for the Turnaround Program. The WBS may not have divided the program into departments and teams in exactly the same way as the Turnaround Team would, but it will probably be close. Also, it provides a valuable common structure for communicating all status and issues between the customer and the Turnaround Program, reducing communication times and communication errors. These savings often will far outweigh the benefits of a more idealistic division of the program organization into teams based on analysis.

If the customer objects to the program task breakdown the Turnaround Lead has selected, customer resentment may linger, and damaging communication errors between the customer and the program organization may occur throughout the life of the Turnaround. It is highly recommended that the Turnaround Lead and their program team derive a program organization that the customer completely understands and is comfortable with, even if compromise is required from the Program.

7.4 Strict Adherence to the Program Plan

In almost all cases, the Program Turnaround Team will write the new Program (or Turnaround) Plan. The content of such plans varies widely but should at least describe the program organization, the functions and responsibilities of each program department, team and key staff roles, schedules, metrics, common program process and protocol used in design reviews, peer reviews, change boards, failure review boards, periodic program meetings, working rules, and ethics reminders. (Refer to Figure 7.6 for the outline of a typical program plan.) It should cover all factors that are of common importance to all program team members, including customer, subcontractors, suppliers, and the sponsoring enterprise. The Program Plan is the plan to save the program and successfully achieve the Turnaround Commitment. Often the new Program Plan will be an amended version of the original Program Plan.

I must now assert something that may appear to contradict my consistent support throughout this book of innovation and continuous improvement. That is, the Program Plan for the save must be followed precisely and consistently, as written. You may ask, how can someone support the rigorous obedience to process when they're trying to improve it all the time? The answer is that you can only reliably improve a process if you clearly know what your current process is.

- Executive Summary
  - Turnaround Commitment statement
  - Description of how to reach commitment
- Organization
  - Organization Chart
  - Role and responsibilities
  - Change management
  - Facilities
  - Project completion
  - Subcontractors and vendors
  - Rules and expectations
- Scope Management
  - Scope Summary
  - Requirements Management
  - Configuration Management
  - Deliverables
- Schedules
  - Master Schedule
  - Schedule Control
- Cost
  - Estimation Process
  - Budget Allocation
  - Budget Control
- Quality
  - Monitoring
  - Control
- Human Resources
  - Acquisition
  - Development/mentoring
- Program Interfaces
  - Stakeholders
  - Reporting and Communication
  - Team interdependencies
  - Metrics Collection
- Risk Management
- Procurement
  - Subcontractors
  - Vendors and services
- Program Information Management
- References (potential)
  - Integration and Test Plan

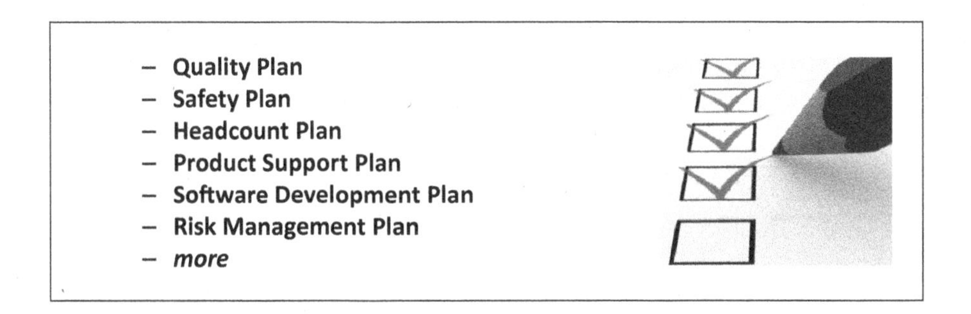

Figure 7.6 This is a typical outline for a Program or Project Plan. The Turnaround Plan is a new Program Plan and should have the same format.

Having the Program understand and consistently follow the current processes documented in the Program Plan is most often more important than having found the best process. Doing so allows reliable metrics and other data on the current processes to be gathered. Using this data provides much more confidence when evaluating how the Program Plan can be improved.

Adhering to the Program Plan allows accurate schedules for future work to be created now, even if they might be shortened with future process improvements. Strictly adhering to the current plan maintains a high degree of successful coordination of the work of the program departments and teams because they are all following the same processes and plan. Strictly following the Program Plan provides a stable process foundation for applying innovations and other improvements in the future.

It then follows that it is essential to have a rapid, easy way for the Program to amend the Program Plan with improvements as program execution is underway. In addition, this amendment process must include a way of quickly informing all program team members of updates to the Plan after they are approved.

7.5 Improvements to Program Plan by Adherence to Procedure

As stated before, the high value to the Turnaround Program of strictly following the Program Plan can only be achieved if there is a formal, yet fast and easy-to-apply way to improve its contents. This must be followed by a process of disseminating changes to all members of the program team in a quick and efficient way—a way that has a likelihood of being quickly reviewed and understood by all program team members.

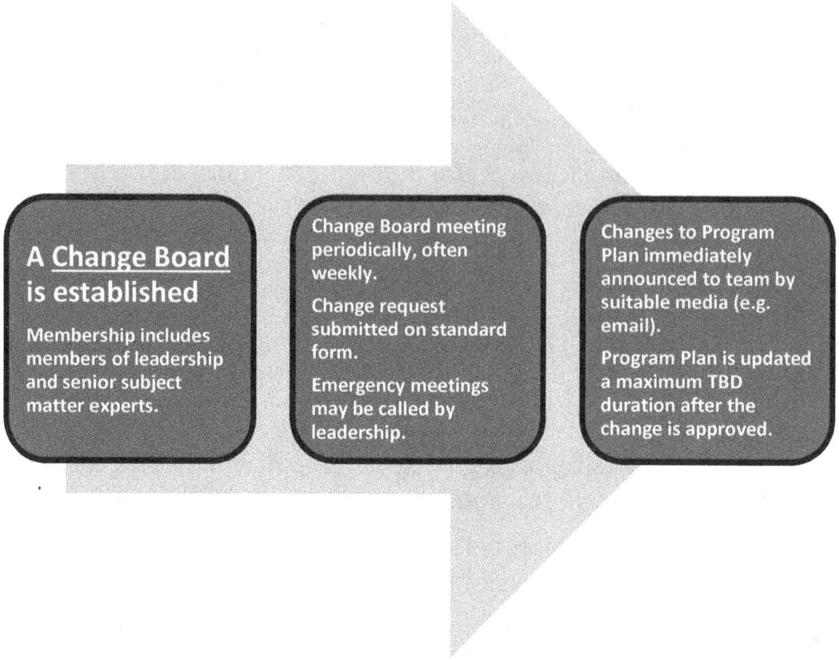

A Change Board is established

Membership includes members of leadership and senior subject matter experts.

Change Board meeting periodically, often weekly.

Change request submitted on standard form.

Emergency meetings may be called by leadership.

Changes to Program Plan immediately announced to team by suitable media (e.g. email).

Program Plan is updated a maximum TBD duration after the change is approved.

Figure 7.7 There must be a stringent process for amending the Turnaround Plan. The program Change Board is often used, with the option to call an emergency review if necessary.

One way that is often used to accomplish this is to convene a Change Control Board (CCB) for the program (see Figure 7.7).

Typically, the CCB is a committee of senior program team members. In its largest form it usually includes the Turnaround Lead or their representative, system-level leadership from the Program, the leadership associated with the area of the proposed changes, and one or more voting members from the customer. The program roles included are often the Chief Scheduler, the Business Lead, the Contracts Lead, the Chief Engineer, and similar roles, depending on the size and subject of the program. The CCB can also include voting members from the other prime-contract team participants (if there are multiple prime contracts reporting to the customer) and members from major subcontractors and suppliers if the program is large.

For a small project the CCB can simply consist of the Turnaround Lead or designee, and possibly a senior product and business leader. What is important is that improvements to the Program Plan must be only made via review a process that is known to the entire Turnaround Team.

The CCB is generally used to review and approve changes to any portion of the Turnaround Plan: product performance requirements, schedules, organization, functions and responsibility of a program team or department, internal or external program interfaces, etc.

I will now suggest a number of guidelines for using a CCB to quickly update the Turnaround Plan. I have applied these with high success on programs that I have managed.

The general CCB meeting is usually convened periodically (often weekly on medium and large programs). The CCB chairperson will have prepared an agenda of proposed changes to be evaluated.

However, sometimes a program team will need approval of a process change that will significantly reduce schedule times and program cost if applied immediately. Therefore, there must be a process included to call a special CCB, sometimes called an *emergency CCB,* which is usually scheduled to convene within 24 hours of the request. This implies that the CCB board members must be "on call." Usually each board member will have one or more backup representatives in the event that they cannot be located. Often the CCB chairperson will facilitate the use of remote participation by the board members (phone/laptop is usually sufficient) for these emergency CCBs.

To insure a complete and accurate review, the person requesting a change from the CCB must provide a written description of the change on a standard form. This form usually asks for:

- A description of the requested change
- The estimated savings in cost, schedule times, error rate, etc. by making the change
- The cost, schedule, and resources needed to make the change
- Predicted risks with implementing the change plus abatement plans
- The consequences of not making the requested change

Usually an approval page is attached at the end of the change request form with places for the signatures of all change board members. Often a provisional approval can be made that includes a clearly written description of the provisional action that must be completed before the rest of the requested change is approved. Also, space is usually allotted for describing why the requested change is denied, if that is the decision.

Each CCB is led by the CCB chairperson. For a small project this chairperson can be the Turnaround Lead or their designee. Each CCB meeting must generate the same artifacts as any program meeting (to be discussed in more detail later). These include a meeting agenda, record of attendance, minutes, action items, decision about proposed change (pending, refused, accepted, amended, provisional acceptance, etc.), and justification if denied.

The CCB chairperson then will often send to all hands a summary notice of what the change is, if one has been approved. Of course, this is to keep all program team members up to date with the current approved content of the Program Plan as soon as a change is made. For a Turnaround Program there should be a requirement that this notice be sent within some fixed time, usually 24 hours, after the CCB has completed its decision.

Then the text of the Program Plan is updated in accordance with the CCB's decision and is available on the program server within some maximum time after the CCB decision was made.

7.6 Each Task Must Have Just One Lead

As discussed extensively in previous chapters, one of the most frequent errors that cause a program to fail is the ownership of program responsibilities being assigned to a team instead of a single individual. When a team is given responsibility for a task, no one person has the exclusive responsibility of making sure they know what must be done, then making sure it is done. As humans, we work best when each responsibility in a work effort is assigned to one person. This person may have a team of individuals supporting them, but they and all the program knows they are 100% responsible for the outcome of the work.

One simple test to verify that each program task is assigned to one individual is during an all-hands program meeting, announce the name of a department or team shown on the organization chart, and ask who is responsible for it. All the people in the meeting should point to one person.

This test will reveal if two very important necessities are met. First, that this Turnaround responsibility belongs to one person, and second, that everyone on the Program knows who that person is. It confirms that the whole program team knows which desk to walk to regarding this responsibility.

In any development program, there is a way of dividing the teams into two categories. There is one category of established, long-term teams that remain intact, usually throughout the life of the program. These are often called *departments* or *product teams* and are usually led by someone who has followed a path of increasingly greater leadership responsibilities. (There can be exceptions to this, especially for a Turnaround effort.) These teams are often responsible for a product or services throughout the life of the program. Examples of such a team's name would be *Cost Accounting, Powertrain,* and *Test.*

The second category of team is usually assembled quickly to perform a special task over a short duration. These teams may not even show up on the program organization chart because of their short lifetime, often spanning from a week to just a few months. The leaders for these teams often have less leadership experience, although not necessarily. This team leadership role can be a good first experience for a team member wanting to move into leadership. First-time leaders should have access to a mentor, as described later. Examples of the names of these teams could be Hinge Fracture Failure Review, District Math Proficiency Report, U.S. Lime Disease meeting, and Version 16.2 Engine Control Firmware Delivery.

There is an endless array of books written about what makes a good team leader. Suffice it to say for a Turnaround Program, any leader should be at least responsible, punctual, a very hard worker, a quick learner, a good listener and communicator, and have a very good understanding of the Turnaround Commitment. In addition, they should have a positive attitude, be known as fair, and be motivated to solve problems, not just work on them.

A task lead should not have a difficult personality to work with or be difficult to communicate with, even if they happen to be exceptionally gifted in their specialty. The overhead of working for a person with communication issues often hampers the fast work pace needed for a program save.

The team leader does not have to be a senior member of the team, nor do they need to be a long-time team member. Recall the lifeboat analogy. If our team leader is credible, fair, and does a good job leading us, let's go. We have work to do!

As pointed out earlier, the highly dynamic leadership roles needed to execute the Turnaround Plan can be excellent assignments to grow potential leaders. This is especially true if mentorship is provided to the aspiring leader. The Turnaround experience can be very valuable for both the leader and their home enterprise.

7.7 Well-Structured Meetings

Well-run meetings are an essential part of a Turnaround Program. Unfortunately, they are very easy to run poorly. Everyone who attended should believe that the results of the meeting were of greater value to the program than the time they invested in preparing, conducting, or attending the meeting. If a team participant does not believe this, then either the meeting was not worth having or there was no need for this participant to attend (see Figure 7.8).

A meeting member who only needs to contribute their input very infrequently may still be of high value to the meeting because of the large impact of their input. But if a meeting member has no involvement in the meeting subject, then their invitation was a mistake. Sending a representative from a program team just to claim the team was represented is a waste of valuable program resources. If, during the meeting, involvement by some team is warranted but there is not a representative from that team, an action item can be recorded.

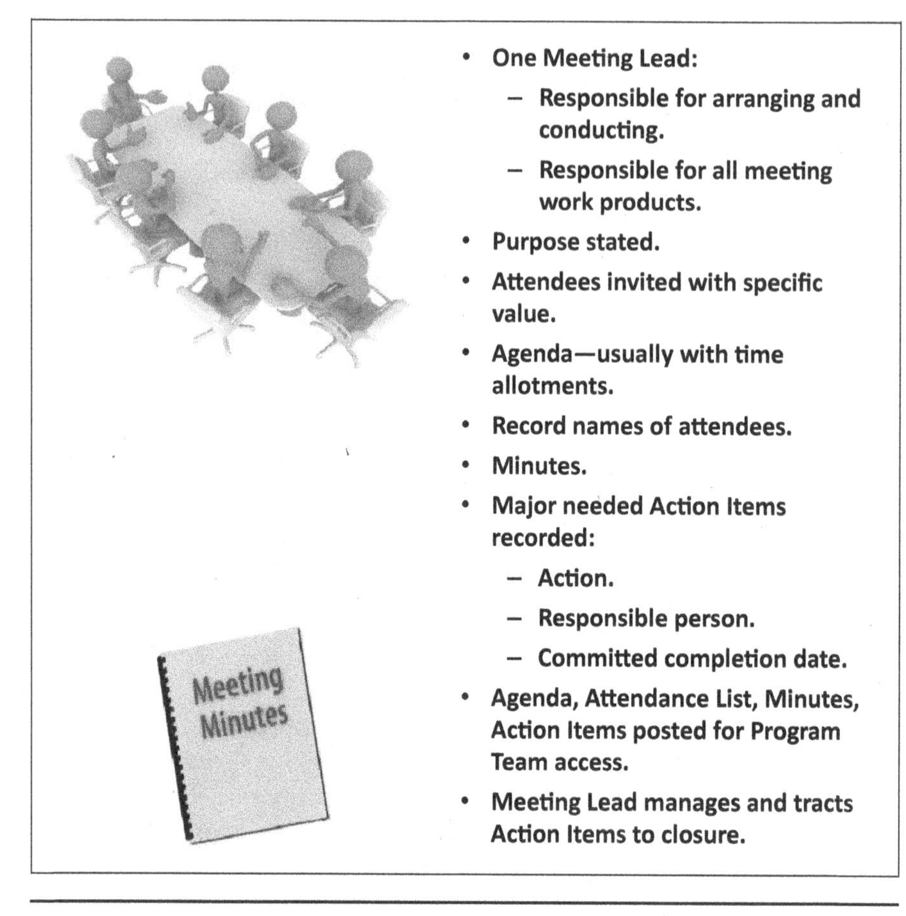

Figure 7.8 Program meetings must follow a structured protocol to be of value. Here is a checklist.

Meeting minutes, including action items, should be available to the Program shortly after the meeting, so there is no need to send a representative from an uninvolved team to just take notes.

Each meeting should address the subject(s) on the agenda and either resolve issues or determine a clear series of actions to resolve them. The agenda should be designed and conducted in a way that provides progress in reaching the Turnaround Commitment. This progress should be clearly worth the total meeting effort invested.

Part of achieving high value from each meeting is to make sure there is a standard process for meetings established in the Program Plan. This process should include the steps to plan and announce a meeting and the artifacts that should be derived from it. The basics this process should describe include:

- **Meeting Lead.** Every meeting that is scheduled and has an agenda must have a meeting lead. This lead makes sure the meeting process in the Program Plan is followed. This process includes how to select meeting participants and announce the meeting, how to start and end the meeting, how to make sure the meeting follows the agenda, how to make sure all required meeting artifacts are generated, etc. The meeting lead must have the power to tactfully postpone discussions during the meeting that are not pertinent to the meeting agenda.
- **Attendance List.** This is unfortunately often overlooked when meetings are conducted in a program. The meeting lead should simply record the names of the people who actually attended the meeting. This of course provides a record of who was involved in making important program decisions, and what teams they represent. This record can be valuable if a program team claims they were not represented when a meeting decision was made.
- **Meeting Agenda.** The Agenda should be prepared and distributed before the meeting to those invited to attend. It should state the purpose of the meeting and list the subjects to be discussed to achieve that goal. This purpose statement is very important and should be concise. It can often be referred to during the meeting to stay on the agenda. In most cases the meeting lead can assign a duration for discussion of each subject. Then the meeting lead can make sure these durations are not exceeded when the meeting is conducted. Of course, time limits can be adjusted if the meeting discussion demands it.
- **Minutes.** The Meeting Lead or their designee should record the minutes of the meeting. This is a written record of what was discussed, by whom, and what was decided during the meeting. The format of the minutes may vary depending on who is recording them, but they should include highlights of discussions that occurred, including who participated; agreements, conclusions, and solutions that were derived; new items discussed that were not on the agenda; and agenda items that were not discussed, if any. The minutes should also include the purpose, time, and place of the next meeting if one is planned.
- **Action Items.** This is a list of important work that must be done to make progress on the decisions that were made during the meeting. Sometimes this work is necessary to gather data to make decisions at a follow-up meeting. Each action item should have a description of what the action is, the name of the single person responsible for completing it, and a due date. This action item list is usually attached to the minutes as the formal record of the meeting. It is the responsibility of the meeting lead or their designee to track each action item to closure.

Action Items must be chosen carefully during a meeting. It is easy to start recording every action the meeting attendees say they will take as an action item. This becomes an excessive and burdensome detail that is unnecessary and will destroy the important credibility of the action items. Maintaining an accurate record of the closure status of so many details will be too expensive to maintain and of little value. After a well-run one-hour meeting, there should typically be five to ten action items; these should be top level, key to the progress of the program save, and achievable in the agreed-upon time.

7.8 KISS, the Three Levels of Problem Solution

The acronym *KISS* (Keep It Simple Stupid) has been around for decades. To this day, successful managers of development programs regarding any subject matter consistently adhere to this mantra (see Figure 7.9). Solutions to a problem can be divided into three levels of value, as follows.

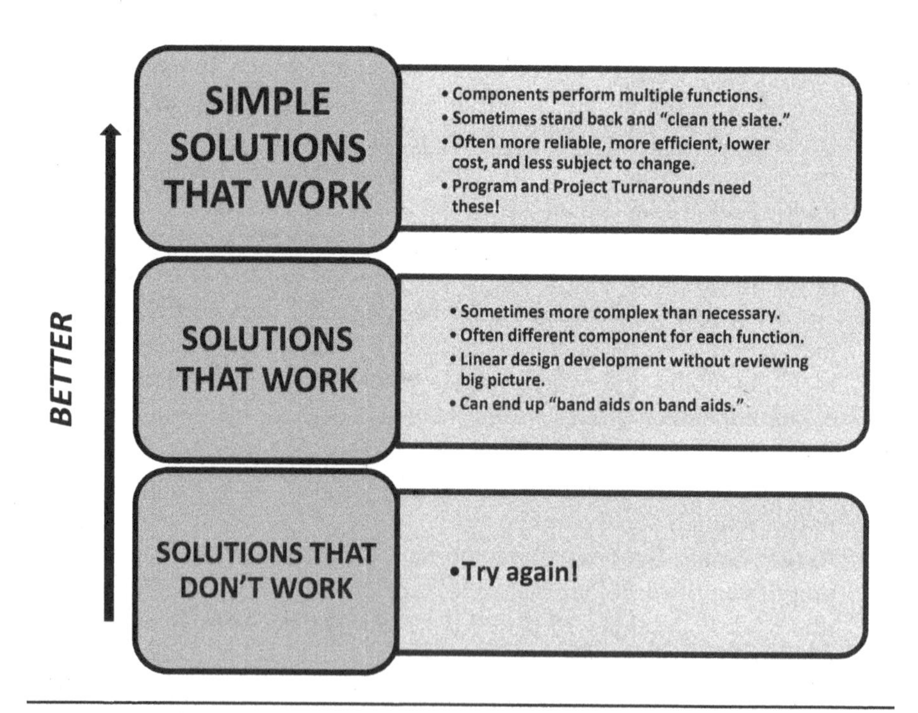

Figure 7.9 KISS (Keep It Simple Stupid) reminds us to develop the simplest solutions. These are the best of the three solution types for a program on step. They often require the most creativity.

7.8.1 The Solution That Does Not Work

This needs very little elaboration. A solution is implemented and it partially or completely does not provide what is needed or solve the problem. In the highly results-oriented environment of a program turnaround, most often a solution that only partially works is considered incomplete. When a development need or problem is solved, it must be solved completely so the Turnaround Plan can proceed confidently.

7.8.2 The Solution That Works

These are of course better than those that don't work but sometimes are unduly complicated. For example, beware of the solution that is a compilation of solutions, each for one apparent symptom associated with the problem. An example of this could be that, given the attendance at the community school is low, the proposed solution is to increase the number of truant officers, send special notices to all the parents, erect posters on the school premises reminding the students of the value of an education, etc. But is this the simplest solution? Watch out for solutions that appear complicated, where each "sub-solution" solves only one "sub-problem." Watch out for a solution that appears to keep applying Band-Aids to the observed symptoms, hoping the problem will go away. At best, the Program Turnaround work may become overwhelmed trying to manage these numerous sub-solutions. It may be much simpler and require fewer resources for the program to solve the root cause, which is often the single cause of the problem.

7.8.3 The Simple Solution That Works

This is the best kind of solution—often, the most intelligent one. An important characteristic of this kind of solution is that each part of it may solve more than one portion of the problem. Maybe the way to solve the truancy problem is to mail a clear reminder to parents of the importance and value of an uninterrupted education, then promise a cash credit each quarter to each family with children who achieve some minimum attendance level. Now you are incentivizing the most persuasive people in every student's life. Their lack of discipline, for whatever reason, is in this example the likely root cause of the truancy. Providing funds directly to incentivize more family discipline instead of hiring more officers and buying a lot of posters may be much simpler and less expensive to manage and justify.

Another example is the ingenious rear seat design in some recent automobiles. These seats perform perfectly as a comfortable seat for all road conditions and for all passengers, yet either the seat backs can be folded forward to establish a large, flat truck area, or the seat bottom can be folded back to expose a high vertical space where the passengers' feet would have been for carrying tall items. Here is an example of an intelligent design solution providing multiple valuable functions, using simple mechanical hinges and latches.

Sometimes while executing the Program Plan, the team will have to stop and approach a solution again with a clean slate to make it simpler. Overly complex solutions often are the result of a team refusing to do this.

Simple successful solutions are very important for a Program Turnaround to succeed. They require minimal effort to implement and maintain, they are easily understood by the whole program team, they operate efficiently and are more reliable, and they allow the Turnaround Program to remain lean and continue to progress at a high rate.

7.9 The Knock at the Door—Innovation Is Here!

In the heat of executing a Program Plan, there will be times when the leadership will appeal to the program team, often in special planning meetings, to invent highly creative solutions for tough problems. For example, "How do we manufacture chips without the low yields we always seem to have?" or "How do we increase per capita donations to our food bank by ten percent without losing subscribers?"

I have had the privilege of conducting many such meetings with brilliant and creative specialists. We constructed matrices listing all the options we could think of and graded them for factors such as benefits, risk, cost, and assigned actions for a follow-up meeting to convene in less than 24 hours. I would return to my office proud that I was able to gather these outstanding professionals and get us started on what I was sure would be an extraordinary solution.

But then sometimes, someone would knock at my door. I would turn to see one of the team members smiling carefully, almost looking embarrassed. Their next words, words I have found to be some of the most important words that can be heard during a development program, are, "I know this sounds crazy, but . . ."

When I hear a team member say this, my world stops and I refuse to be distracted by anything else. I have observed that many if not most innovative breakthroughs come from individuals, not teams. They are based on an individual inspiration that can come any time in a person's day (or night). The idea that energy and mass are equivalent or that rocket ships can propel payloads much farther with staging each came from the mind of one individual. These were "crazy ideas" that were evaluated, validated, and highly valued later on.

I keep pencil and paper on my nightstand in the event that I have one of those ideas during the night. Others have told me that their idea occurred to them while eating breakfast, driving home, reading the paper, or sitting in front of the computer. It can happen at any time. The challenge to the project or program leader is that they cannot force these inspired ideas to happen, but they can encourage them and facilitate a thorough evaluation of each one.

As a leader in a Turnaround Program, keep your door open to these ideas all the time. They can come from any team member at any seniority level. They can pertain to any part of the Turnaround, from reducing test failures to ordering office supplies.

Never, ever show disappointment when being offered an innovative idea, even if it initially appears not to be feasible. Listen to whole idea carefully. In fact, try to be an advocate of the idea, even if it at first sounds like it probably will not be used. This will encourage the person presenting the idea to keep innovating, even if it turns out that their current proposal is not feasible. Give each suggested idea the same due process, with subsequent reviews by team specialists if it is potentially useful.

It is important that an individual be complemented for proposing an innovation or improvement idea, even if it is not used. Sharing an attempt to improve the program with the Turnaround Team should only be limited by the desire for privacy expressed by the person you wish to recognize. Team members who come forward with a so-called crazy idea that eventually proves to be highly beneficial to the program should receive special recognition in front of the program team. The resulting quality improvements, cost savings, schedule savings, etc. should be quantified if possible and shared. All members of the program team should know that any suggested innovation they provide is highly welcome, even if ultimately it is not used. They should also know that the attempt to put forward an idea they believe will improve the program will positively affect the evaluation of their work performance.

7.10 Team Rhythm

If you put ten people in a large rowboat, give each an oar, and ask them to row across the river as fast as they can, pandemonium may break out. With each individual rowing as fast as they can, oars will likely bump into each other, the boat may weave left and right, and there may be a lot of water just wastefully splashing around.

But if you put an eleventh person onboard and ask them to yell out "stroke" at a consistent interval and instruct the rowers to pull their oars when commanded, magic happens. The boat will go in one direction and reach a high speed. All the energy of the ten rowers is now being usefully applied to propelling the boat (see Figure 7.10).

What's the difference? At least two things. One, of course, is that there is now a leader on the boat. But two, the team is following a rhythm.

Rhythm is often overlooked when managing development programs. Some managers may ask if it is more efficient to perform tasks on just an "as needed" basis. Isn't program rhythm more appropriate for a steady-state enterprise, such as providing a fixed service or manufacturing goods? Is program rhythm of value for an intense turnaround program containing tasks that are constantly changing?

- **Apply common formats for planning and tracking, including:**
  - **Cost.**
  - **Schedule completion.**
  - **Achievement of required performances.**
  - **Risk management.**
- **Rhythm Examples**
  - **Daily project status meeting:**
    - **Progress to plan.**
    - **Issues and workaround plans.**
    - **Risk and avoidance/ abatement.**
    - **Help needed.**
  - **Weekly meetings:**
    - **Failure Review closure.**
    - **Change Control Board.**
    - **Program or Project report to enterprise.**
  - **Periodic peer reviews.**

Figure 7.10 The Turnaround Team must work to a common "beat" to be most effective. This applies to conducting periodic meetings and applying common formats for exchanging information.

We should remind ourselves that even in the heat of a program save, we humans still need periodic information exchanges. This is necessary to maintain a uniform team understanding of the current status of the program and the detailed plan to get to the next milestone. This program rhythm is necessary for a high level of progress. In addition, it can provide each team member with an accurate reference to evaluate their own performance.

The Turnaround Program must have daily and/or weekly team review meetings to provide the necessary rhythm. Each program meeting (as discussed earlier) should have an agenda, be concise, and carry a list of important action items. Only those team members who participate in the information exchanged should be invited to the meeting. In some instances, this can be a sizable portion of the workers; for example, if the program turnaround leadership conducts a monthly all-hands status review, the entire team would be invited.

Some typical examples of periodic meetings applied in turnaround programs include:

- Daily Progress to Plan Meeting. Large, complicated turnaround programs have a daily meeting to review whether work scheduled for the preceding day has been accomplished. This meeting is usually conducted first thing in the morning. The scheduled completion milestones for that day are also reviewed and established. Usually, at a minimum, those leads reporting to the Turnaround Lead are expected to attend. In addition, issues with workaround plans are discussed. Requests for more resources, special assistance, facilities, and other necessary items are made by any of the attendees at this meeting. These requests can be made for assistance from inside or outside the program.
- Business Meeting. All team leads and/or departments leads, if they exist, attend with their business specialist to review performance to cost and schedule commitments. These meetings are usually conducted weekly or biweekly. Usually team/department progress highlights to date and the work planned to be performed over the next reporting period are shown by each program team or department. Special issues, new risks, and workaround plans are also presented. This meeting also provides another opportunity to ask for help from sources internal or external to the program via the program leadership. The use of common program-tracking methods and formats for schedules, cost accounting, achieving committed performances, identifying and managing risks, etc. are an additional help to make sure everyone in the Turnaround Program is "pulling their oar" the same way and at the same time.
- Change Control Board and Failure Review Board meetings are usually conducted once a week when applicable. There should be a special

provision in the Turnaround Plan to call a special meeting of either board if an urgent issue requires it.

- Most of the program departments or program task teams will have periodic status reviews at which the performance of their team is reviewed and flow down of information from the enterprise is provided. These are often conducted once a week.
- In a few cases, periodic peer reviews will be conducted of products being developed within a program department or team.

7.11 Plan to Find and Correct Product Errors Early in the Development Flow

When planning a production program in which many copies of the same product are being made after unit testing and system development testing have been successfully completed, the program can remove some of the early testing steps. With a qualified and strictly controlled manufacturing process, the program can assume that sufficient unit-level product quality has been achieved and can justify just performing system-level testing.

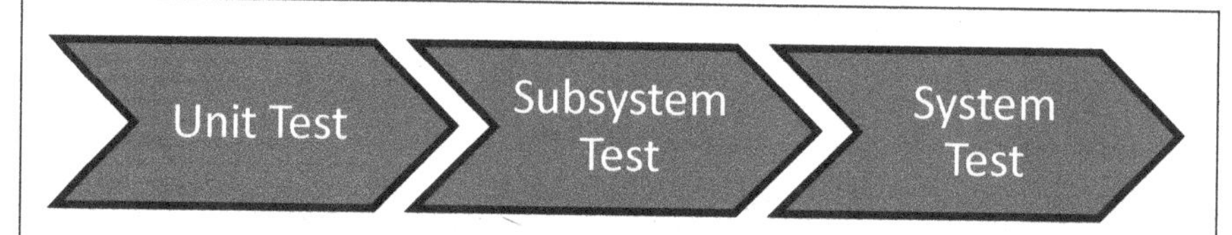

- **Cost of correcting grows exponentially with each test step.**
  - **More neighboring components in danger of being damaged.**
  - **More costly resources and facilities waiting for correction.**
  - **Less opportunity for other task work while correcting.**
  - **True for all subject matter, plans, processes being implemented.**
  - **Cost/schedule pressure to reduce regression testing, may add risk.**
  - **More attention from customer and enterprise!**
- **Insight from early failures improves test coverage and lowers cost.**
- **When root cause is determined, augment the test plan to find the failure or fault as early as possible in the future.**

Figure 7.11 The review or test flow must be designed to find product issues and faults as soon as possible. This greatly reduces total cost and schedule time lost for the Turnaround Program.

But for a development program, there have often not been enough products created to validate the manufacturing line and eliminate early testing. Therefore, early testing must be performed to find any defects in the newly developed product as soon as possible. This is also true for a product that may have been manufactured for a long time but has had a design change.

The cost of fixing a test failure increases roughly exponentially with how late in the development plan the fault is found (see Figure 7.11).

For example, it is much less expensive to correct a fault in a single electronic component or a single line of software code during early bench testing. The cost of the fix escalates greatly if it is found in unit testing, more if found during subsystem testing, and even more during product acceptance testing. Why is this?

- The number of program team members and enterprise facilities impacted by finding a fault increases greatly as the product enters higher levels of integrated testing. Compare, for example, finding that the potting on a thin-filmed capacitor is defective during bench testing instead of subsystem testing. In the first instance, perhaps the design engineer, a technician, and possibly $150,000 of enterprise test assets are not productive while this defect is being corrected. This represents some loss of program cost and schedule time. In box testing it might be more like five to ten engineers, technicians, and facilities experts, and a couple million dollars of enterprise test assets that are nonproductive while the same fault is being corrective. During product acceptance testing, this same defect could render fifty people nonproductive while correcting the failure, wasting millions of dollars.
- When a failure is found when the product is more integrated, it is harder to isolate the cause and verify the correction. This adds more to the cost and schedule time needed to correct the defect.
- When a failure is found at a higher level of integration, there is usually less opportunity for other parts of the program to perform useful work while waiting for the correction to be made. This is partly because these other parts will spend extra time planning and executing special regression tests to verify that their part of the system works after the correction to the fault is made.
- Even with well-planned regression tests, the other functions in the product will likely not receive the same test coverage when a fault is found and corrected late in the test sequence, as opposed to if it were found earlier, before product integration. This increases the chances of a new undetected fault resulting from the previous fault correction remaining somewhere in the system.

- Failures encountered during higher levels of product integration get much more attention from enterprise executives and the customer! Often one or both organizations will provide special teams to assist with correcting the failure. Help from these sources adds extra process, which in turn adds cost and schedule delay. In addition, receiving help from the enterprise and/or customer is usually embarrassing for the Turnaround Lead and may even adversely affect the evaluation of their work performance.

Again, the series of tests in the Program Plan must be designed to find product defects as soon in the series as possible. This applies to any work of a Turnaround Program, including developing a new process, a new plan, or a new product.

Unit tests should be formal even if they are just performed by the designer/developer of the component being tested. The method for performing and documenting unit tests should be stipulated in the Turnaround Plan. Peer review of unit test plans is highly recommended.

These formal unit test actions may seem excessive to those whose instinct is to start integrating as soon as possible. But in fact this investment in effort will pay back many times over in preventing cost overruns and missed critical milestones. A well-engineered test plan is often a major differentiator between a program that is falling seriously behind and one that is successfully making progress.

Keep in mind, if the program leadership suspects that they are spending too much time and money on unit testing, they should collect and examine metrics on what product defects are being found by this testing. If there are certain unit tests that have never failed, there might be justification to remove these tests. However, being able to justify this removal is rare. Instead, gaps in unit test coverage are more frequently identified. If this is found, the test coverage of existing unit tests should be modified or more unit tests should be added.

A proven top-level method for finding the root cause of a test failure, quickly and accurately, is discussed later.

7.12 Watch the Flank!

I have worked on complex develop programs that managed their daily work priorities by following the tasks on the critical schedule path. This path is the series of scheduled work steps that takes the longest of all such work paths for the whole program. The theory is that the only way to shorten the total program schedule duration is to shorten the critical path. So the work steps on this path are given priority.

Of course, when the original critical path is successfully shortened, another series of interdependent work steps becomes the critical path. And here is the important part: the development program may have paid no attention to work steps on this new critical path, so the focus of the program has to abruptly shift to a new series of problems. If more attention had been paid earlier to the tasks on this new critical path, the time to program completion may have been made shorter—a cost and schedule reduction opportunity lost.

Also, having to abruptly shift from concentrating on one set of work steps to another results in confusion, errors, and the inefficient use of time. Therefore, program turnaround activity will complete the Turnaround Commitment sooner if attention is paid not only to the work steps on the current critical path, but also to those steps on work paths that would become the critical path if the current one were shortened.

Today's computer-based scheduling tools allow complex development schedules to be created quickly and with high accuracy. One time-saving capability is to display a schedule in different formats. As introduced in Chapter 5, one such display is the PERT format. This format is highly intuitive. It shows a complex program schedule as a series of task symbols such as bubbles, each representing a development step connected by lines representing the interdependences. The schedule program can instantly highlight the series of bubbles and lines that represents the longest path of schedule dependencies—the critical path.

But there are tasks on other development paths that are scheduled to be completed at the same times as those tasks on the critical path. These other tasks must be completed on time as well. The emphasis they get depends on how much more urgent the tasks on the critical path are and how much longer they will take. Nevertheless, some attention must be paid to these development paths "on the flank" to the critical path to make maximum use of program resources. This allows faster transition to the new critical path if the original one is reduced in duration, and it will allow the Program Turnaround Team to achieve the Turnaround Commitment as soon as possible (see Figure 7.12).

Displaying the schedule in other formats can help identify those critical work tasks that need to be completed in parallel with current critical development tasks. PERT displays, for instance, can be commanded to show the second, third, etc. critical paths in the Turnaround Schedule.

In addition, the more traditional GANT format can be used. It lists the major program schedule tasks with indented subtasks shown underneath them. Schedule start and stop milestones are shown with horizontal schedule bars on the right. Scanning down these schedule bars vertically will easily show all the tasks that were planned to be worked on in a given day.

Another valuable source for warnings of program risks that are developing "on the flank" are the concerns brought forward from the Turnaround Team members. The folks working the tasks full time will often detect a serious issue on a parallel program development path before program leadership does. The

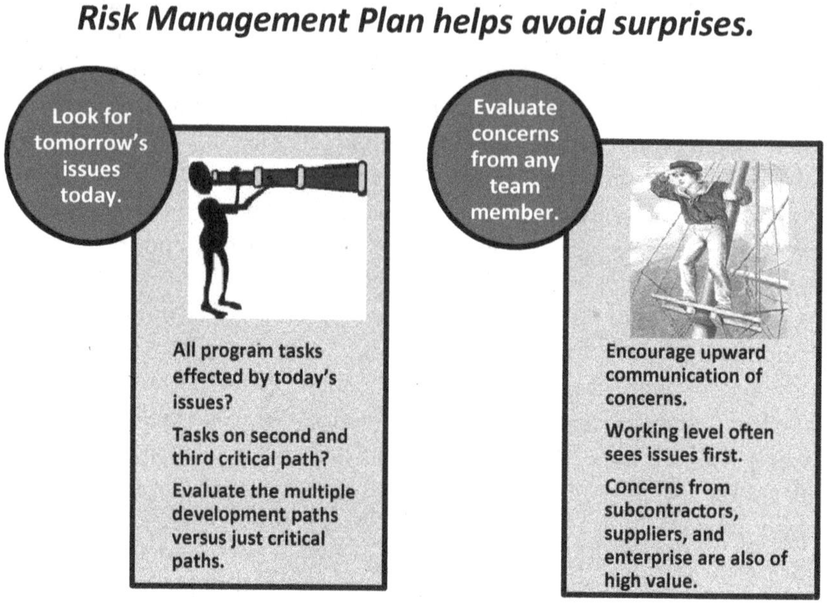

Risk Management Plan helps avoid surprises.

Look for tomorrow's issues today.

All program tasks effected by today's issues?

Tasks on second and third critical path?

Evaluate the multiple development paths versus just critical paths.

Evaluate concerns from any team member.

Encourage upward communication of concerns.

Working level often sees issues first.

Concerns from subcontractors, suppliers, and enterprise are also of high value.

Figure 7.12 The program leadership team must completely evaluate any concern, from any team member, about an unknown risk. Look out for risks not on the current critical path.

good leader in a program turnaround will always listen intently to critique, worry, or a suggestion from a team member. Never dismiss a concern voiced by a team member about a development path not on the current critical path!

The four major performance metrics to track for all programs are adherence to requirements, adherence to cost plan, adherence to schedule commitments, and risk mitigation.

Risk mitigation requires a Risk Management Plan, which is often underemphasized and even overlooked in some programs, perhaps because maintaining this plan requires carefully determining, grading, and documenting developments that can endanger the success of the program. This analysis directly grates on the teams' optimism about doing well in a new development program. It can scare members of and supporters of the Turnaround Program at the start of executing the Turnaround Plan.

Yet, many times a program team will work feverishly to complete a task, knowing that there is a big issue that must be addressed to be successful, and for some reason, no one seems to want to worry about it now. Is the program leadership so overwhelmed with development issues that they feel justified in refraining from comprehending the reality of the big picture?

For example, I was assigned to help a team doing an excellent job of building two of three required subsystems, knowing that the third subsystem had to be completed to deliver the system. All program leadership knew that the real date and commitment for delivering the third subsystem was still unknown. For some reason, they refused to acknowledge the catastrophe ahead if the third subsystem was not completed. The solution was simple: adding focus on completing the third subsystem was sufficient to turn the program around.

One of the benefits of following a Risk Management Plan is avoiding this trap. The risk management team usually consists of representatives of each of the subsystems or elements of the product being developed. This team identifies the most likely risks for the program. They grade these risks, often as a product of the likelihood of the risk becoming an issue times the consequence of the issue. They often develop and monitor an abatement plan for each risk that is graded above a certain value. This plan includes a set of actions to avoid the risk as well as a plan to mitigate the issue if it happens. The typical risk management documentation for a program will be discussed in Chapter 19, Document and Follow.

Executing a Risk Management Plan is a major factor in watching the flank. It is an essential process that any program must exercise to track and mitigate all the risks that may cause the program to fail. It is necessary to achieve the Turnaround Commitment.

7.13 Hiring Rules

Some managers make the serous error of treating "headcount" like a commodity. This is a serious mistake! The term "good people" is touted commonly in all industries—but what makes a team member "good"?

Remember, any development program is an exercise in achieving high forward speed. In the case of a program turnaround, the team wants to completely achieve its Turnaround Commitment as soon as possible. The program leads can only do so much to identify the shortest path to completion and assign the work to the teams in the most effective way to accomplish this. Once the team members receive their assignments, it is up to them to plan their work in such a way that they achieve or exceed the required product quality in the shortest time possible. For a team member to do this requires more than education and experience—it also requires a special attitude.

7.13.1 "If I Could Only Ask One Interview Question"

If you were interviewing a candidate to join your Turnaround Program team and you could ask them only one question, what would it be? "What was your

GPA in college?" "How many years of work experience do you have?" "Do you like working with a team?"

Here is my question—for me, often the most important question I ask during an interview:

"Tell me about something you have done that you thought was really extraordinary."

What I am interested in hearing is not the content of the answer but its format. Almost always the response will fall into one of two categories: the "good answer" and the "great answer."

The good answer: "Oh, I really enjoyed working on the Alpha program. I loved everyone on that team. We got along and made good progress on it. My boss was a good listener, and we actually used some of the ideas I had. I was sad when that program ended. We had a great celebration party at the end. I think the work we did was good. I would love to work on a program like that again."

The great answer: "The Alpha program was really interesting. I started to work the packaging design but saw that there might be a way to reduce the manufacturing cost. So I talked to a few of the other team members, and we put together a plan that we thought could reduce the cost by up to eleven percent per unit. We went to the boss and convinced him to let us try it on ten units. So five of us worked our tails off to make this work. Like I hardly went home for two weeks. And we killed it! We actually reduced the per-unit cost by thirteen percent and moved the program completion date up by five and a half weeks! I don't think anyone thought we could do it!"

Please stand back. Note that the good answer person sounds like they mostly thrive in what I call "problem space," the great answer person seems to thrive in "solution space."

The good answer person enjoys the harmony of a good team making good progress. The most memorable part of their work experience sounds like the relationships they shared with their coworkers and their boss. They seem to imply that completing the program was something that would happen in due time, supported by a professional and competent team. But it did not sound like this was a major factor in their mind. They even seemed a little disappointed when the experience came to an end.

Now let's dissect the great answer.

"The Alpha program was really interesting *(curious mind).* I started to work the packaging design but saw that there might be a way to reduce the manufacturing cost *(innovative, on the lookout for a better way).* So I talked to a few of the other team members, and we put together a plan that we thought could reduce the cost by up to eleven percent per unit *(team worker, quantitative improvement estimate).* We went to the boss and convinced him to let us try it on ten units *(followed protocol, sold the idea).* So five of us worked our tails off to make this work *(teamwork, hard work).* Like I hardly went home for two weeks *(will do what it takes).* And we killed it! *(passionate).* We actually reduced the per-unit cost by thirteen percent and moved the program completion date up by five and a half weeks! *(measured real improvement against estimated, glad to get the program done).* I don't think anyone thought we could do it! *(will take a risk).*"

The great answer person is entrepreneurial. They want the program to solve the required problem but they want it to have high business value to the enterprise as well. They appreciate the bigger business picture. They spontaneously pull teams together to solve problems. They will take risks and get gigantic satisfaction in doing what they thought they could do. They seem to be very happy to get program completion behind them so they can move on to the next challenge.

So who would you like on your team? Of course, not all the team members can be the great answer type. All excellent teams, including turnaround teams, need the "Steady Eddies" (good answer type) that do a highly professional job completing the work exactly as requested, on time. But you will need a certain number of solution-minded innovators to inspire others and achieve a successful program save.

7.13.2 Key Attributes

So let's review what you're looking for in those great answer professionals who are considering working on your Turnaround Program.

- **Useful knowledge of their specialty.** More than just academic experience, but also solving numerous real problems with actual applications.
- **General knowledge of the whole product.** Have knowledge of the planned operation, functions, characteristics, and needs of the whole product, not just the subject area they will contribute in.
- **Innovator.** Always looking for ways to improve their work. This includes increasing quality, reducing cost, making the solution simpler, improving performances, etc.
- **Team Builder.** Pulls together team members with the sole intent of getting the task done.
- **Obsessed with getting problems solved.** Uncomfortable when problem is not solved.

- **Proud of measured results.** Proud not only of completing a task but of the resulting improvements in the measured values for cost, schedule, performance, reliability, etc.
- **Not too concerned about having a next assignment.** Focused on solving the problem at hand. Not worried they are going to put themselves out of a job.

There is no stereotypical image of the great answer person. They come in all ages, ethnicities, physical status, religions, personality types, sexual orientations, etc. Listen carefully for the great answers! You will need these people to achieve the Turnaround Commitment.

7.13.3 Other Important Questions

In addition to the important interview question discussed above, there are other questions that are important to ask when interviewing a professional to support a Turnaround effort.

- Ask questions regarding the specific technical expertise of the interviewee. This requires the interviewer to be somewhat knowledgeable about these technical elements. There are two reasons why these kind of questions are important.
    - First, this determines if the applicant has a high level of knowledge and experience in the expertise you are trying to hire. For example, "Have you ever used cost-volume-profit analysis to establish sales price?" "Have you ever built a software operating system?" "Was the operating system time driven or task driven?" "What accounting assumptions did you use to perform your last municipal audit?" "Have you used finite elements modeling to perform stress analysis of a metal part?" "Have you reduced operational overhead at a nonprofit?" "By how much?" "Did you use computer tools to track the savings?" The point is that you're not just interested in what school they went to and their experience level, but whether they really understand their expertise. Are they current with the state of the art of their expertise? Are they interested and even enthusiastic talking about what their expertise is? I have had to dismiss applicants who had technical degrees from name schools with high GPAs but could not explain how they would approach solving a rudimentary problem in their field!
    - Second, these questions can pose a challenge to the applicant to be good enough to be hired by the Program. It has amazed me how many good applicants who feel their knowledge may not be up to the high

standards of the job you are representing become very intent on being hired by you! Perhaps they feel challenged to show you they can really do the work. Regardless, this is as far better relationship than one of trying to talk an applicant into working for the enterprise.

- Ask how they work with a weak team member. There is no completely right answer to this question, but there are wrong ones.

  There is one instance of a wrong answer I will never forget. I was interviewing an extraordinary engineering graduate with two PhDs from name schools. This person's command of the technical material was excellent. But when I asked about working with weaker team members, this person loudly blurted with passion, "I won't work with any turkeys!" Of course, every team has strong and weak members. This person did not get the job.

  Any development program, and especially a turnaround program, is an intense team effort. The ability of each team member to successfully work with the rest of the program team—which of course will consist of individuals with a wide range of talent, experience, and capability levels—is critical.

  One of many good answers to this question is, "I will work with the weak team member to help find the best way they can support the team. Everyone has their strengths."
- Are they willing to work extra hours? There is nothing wrong for a worker to insist they limit their workweek to 40 or 45 hours. There are many good reasons for this, including spending time with the family, preserving a balanced lifestyle, attending to health-based limits, etc. These are legitimate limits that must be respected. Unfortunately, during the struggle of turning a failing program around, limits on the work hours needed cannot be guaranteed.

  A discussed earlier, progress in a turnaround program is not measured by work hours spent but by the completion of the needed tasks. Team members on this kind of program must be willing and ready to put in the extra hours necessary to complete their assigned work on time. With unforeseen problems occurring, a task will often require more time to accomplish than originally estimated. Most professionals appreciate this uncertainty, and they and their families are willing to make the sacrifices to complete the work on time.

  Ask this question during the interview. Some candidates may surprise you with their insistence on not working extra hours for any reason. If the candidate is not willing to make time sacrifices when necessary, they are probably not a good selection to help save a program.
- "When can you start working for us?" This question seems so incidental. Why bother discussing it? Yet, the answer might surprise you and may even be a cause to not make an offer.

The applicant may not be sure they want to move across the country to work with your enterprise. Maybe they haven't told their current boss they're thinking of changing jobs and do not know when their release date will be. Maybe the applicant is just too afraid to tell their boss they're going to quit. Maybe they want to interview with some other companies before they make a decision. The list of responses that avoids a commitment to start supporting the program turnaround is long.

You may have spent three or more hours interviewing this candidate and giving them a tour of the work facility. And now the candidate is telling you they are not even close to making a decision to work on your program? My recommendation is that you cut your losses. Appreciate that if you had not asked this question you might have wasted much more time with emails and phone calls finding out this prospective employee does not plan to accept your offer. Some might surmise this may be a nice way for the candidate to say they don't want to work for your program. But if that's the case they should simple state that and not disguise it with wasteful indecision. Professionals working closely together to save a program must be honest and direct about important decisions. Having your job offer refused by indecision may be a blessing.

- Do not be personal. Do not delve into any discussion of the personal aspects of the applicant's life, even if it seems to be an innocent step toward developing a more candid exchange. This is not just to avoid the obvious litigation that can result from a personal question. As a representative of a troubled program needing help, as well as of the sponsoring enterprise, you want to clearly signal that you are only interested in the applicant's professional performance. You might ask a coworker how their children are or how the golf game went, but not a job applicant.

Some applicants choose to add a "Personal" paragraph to their resume in which they discuss what they like to do in their pastime. Ignore it! Marital status, family size, country of origin, age, hobbies, religion, ideology, anything personal must be off the table during the interview. If the job candidate has a physical disability that might be an issue with performing the needed work, a human resources representative or equivalent person in the enterprise should be called to evaluate it. If the turnaround is in a small company or start up, perhaps an employment attorney should be hired to assist with the evaluation of accommodations needed for a disability.

Discussing the personal activities of the applicant may not be just illegal and unprofessional, it may actually bias your decision to make a job offer. As unbiased as we believe we are, answers to personal questions may wrongly color your opinion of the applicant because of your own personal feelings. For example, "He did okay on the interview, but she's a licensed pilot. So am I. I think all pilots are special." Or, "The interview went very well, but he's never been married. Why is that?" These examples have no relevance to your decision to hire.

I repeat: do not discuss any personal subjects with the applicant! Do not allow your applicant to start such a discussion. You must base your hiring decision purely on the applicant's ability and attitude regarding the job you need done.

7.13.4 Quickly Hiring Many

Some program saves may a need to recruit and hire a large number of highly capable team members in one or more specialties in a short amount of time. After all, part of the reason the program was originally failing may be due to insufficient resources in a certain area.

For this situation, it is tempting to enlist a hiring firm or run a search on an online website to find team members. But as demonstrated above, a careful personal evaluation must be made, often by the leader of the task area needing the added team members. One way to accomplish this that I have found to be very effective is to temporarily assemble what I will refer to as a "recruiting funnel" (see Figure 7.13).

The first step is to identify a large source of applicants to consider. It can be a job fair, a commercial database, a company database, a database of recent applicants for jobs in the enterprise, or other resource.

Then select some number, usually three to ten, from your existing team to be screeners. You must briefly educate these people as to the details you are looking for, including education, experience, teamwork ability, innovator history, etc. These people become the "funnel." Based on your requirements, they will screen and select the applicants they think might be qualified, then send them to you for an interview.

This works surprisingly well! At one job fair I was able to make spot offers to four excellent senior professionals with unique specialties in just three and a half hours. Three of the four accepted the offer on the spot, the fourth accepted a few weeks later. They were major contributors toward turning the troubled program around and performed excellent work for the enterprise years after they were hired.

7.14 Subcontractors Are Team Members—Nothing Less and Nothing More!

This subject is described in more detail in Chapter 9, Contract Success, but it is worth mentioning it here as well.

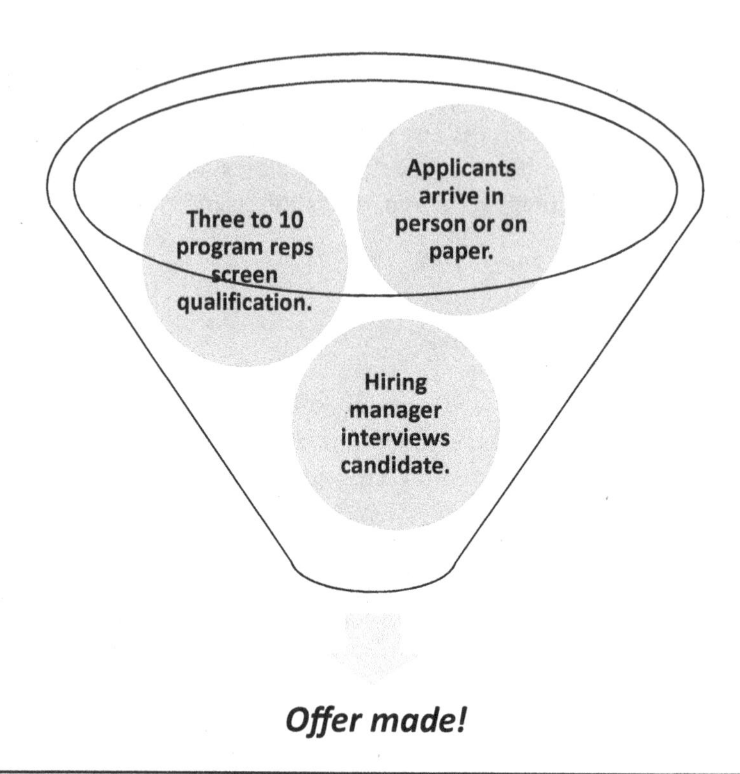

Figure 7.13 A recruiting team (funnel) to quickly select the best job candidates from a database or job fair can be assembled. It assures finding high-quality candidates. If possible, job offers at recruiting events should be made the same day.

Good contractors supporting a program save are a godsend, but a wrong relationship with a subcontractor can be expensive for both parties and can even be one of the causes for a program to fall into trouble.

I have observed two extremes in managing subcontractors. The programs that had one or the other of these extremes did alright, but did not operate at their peak level. For a program turnaround, either one of these extremes can prevent a program in trouble from being saved. I will refer to the relationships with these contractors as the "Shrewd Prime" and the "Obedient Prime."

7.14.1 The Shrewd Prime

This prime contractor extensively scrutinizes every proposal and contract letter from their subcontractors. These primes constantly negotiate for lower subcontract prices and increased delivery benefits. This prime monitors each phase of each subcontract with intense daily scrutiny. There is no question who the boss is—it's the prime! There almost appears to be an underlying assumption that the best product is a result of constant debate and negotiation between prime and subcontractor. The prime completely dominates the management of the subcontract. The prime may not acknowledge that the subcontractor has special capabilities and expertise the prime does not have.

7.14.2 The Obedient Prime

This often starts with the prime contractor choosing a noted subcontractor who has completed a similar job for many other programs or projects. Often the customer will consider this subcontractor the best selection and possibly the only one who should be considered for the job. The customer may have had a history of working successfully with this subcontractor. The prime may even allow this subcontractor's personnel to occupy part of their facility. The subcontractor often may independently lead their work and report their status only periodically to the prime. Reviews of the progress to plan of this subcontractor may be infrequent. Direction from the prime may be limited to "management by action item." There is often very little daily direction given to the subcontract team by the prime's management team. If the subcontractor has a disagreement with the prime, they might contact the customer directly, often via customer contacts developed in the past, to influence the outcome.

It is important to note that neither of these prime contractor–subcontractor relationships are healthy. Either can cause the Turnaround Program to fail.

In the first case there is relentless domination over the subcontractor by the prime. Unfortunately, the natural friction that will result stymies the high value the subcontractor should bring as a program team member. The prime should have hired this subcontractor because they are able to do something that the prime contractor cannot do. With this intensive scrutiny, the subcontractor will do a good job, but they may not be allowed enough self-direction to do a brilliant job. It will unfortunately be less likely that this subcontractor will suggest a highly valuable idea for the program that the prime does not have the expertise to think of.

In the second case the subcontractor has virtually been allowed to invade part of the prime contractor's work. Recall the earlier imperative that each program responsibility must be assigned to one individual. In this case, the responsibility of successfully accomplishing the program ends up being the responsibility of two individuals—the Turnaround Lead and the manager working for the subcontract. In its worst manifestation, the subcontractor communicates directly with the customer. This result is at the very least a cumbersome and expensive way to run a program with multiple interfaces to the customer. It may be impossible to recover a development program in trouble with this relationship between the prime and the subcontractor.

7.14.3 Good Subcontract Management Guidelines

So how does a program turnaround manage a subcontractor so they are a highly valued team member? Here are some guidelines:

- Establish the basic subcontract management team at the prime. As discussed in detail in another part of this book, there are three positions that must be assigned for each subcontract.
    - **Subcontract Manager.** This person leads the entire subcontract and is the point of contact in the program for the subcontractor.
    - **Subcontract Administrator.** This person works with the Subcontract Manager to make sure all government statute and enterprise contract standards are followed. They oversee the evaluation of the subcontractor candidates and adherence to the selection process to verify completeness. They make sure all required documents that establish and define the contract have been created and provided by the prime and the subcontractor. They help verify that all contract-required deliverables are provided with adequate quality and on time by both prime contractor and subcontractor.
    - **Technical Lead.** This person also reports to the Subcontract Manager. They thoroughly understand all subcontract requirements and the design and/or organization of what the subcontractor provides in response to these requirements. This person is the technical expert in the program for understanding and explaining what the product is that the subcontractor is providing. They are responsible for making sure that planned design reviews and testing by the subcontractor are adequate. They must verify that the subcontractor has achieved all technical requirements via acceptance test, analysis, similar past work, etc.
- Treat all subcontractors fairly and with the same enterprise process and/or turnaround process documented in the Turnaround Plan. Each subcontract management team must provide the freedom for each subcontractor to suggest innovations and process improvements to the prime. Providing credible innovation and improvement recommendations in earnest, even if they are not entirely incorporated, should have a positive effect on the subcontractor's award fee, if this fee is part of their subcontract.

- The prime should respect the subcontractor as an equal. This attitude is necessary to open the door to special ideas and innovation from the subcontractor. The good subcontractor does not need to be reminded that they are subordinate to the prime. If the prime constantly treats the subcontractor as a subordinate, the subcontractor will do exactly what they have agreed to do and usually nothing more. This absence of creative improvement ideas from the subcontractor can be a tragic loss, especially for a program trying to do everything possible to get back on track.

  Within the constraints of proprietary data and/or competition-sensitive information, the subcontractors should be asked to be members of the product teams they are supporting. They should be invited to program reviews. They ideally should be invited to team meetings that will be planning the work they are supporting.

  The subcontractors must have signed the necessary nondisclosure agreements. If you're afraid that a subcontractor might use program planning and status information against the prime, you've probably hired the wrong subcontractor. Conversely, many times a good subcontractor can actually help the prime get through an "embarrassing" situation known just to the subcontractor and the prime. They may have been there once too.
- A candidate subcontractor should not be led to feel favored because they have worked with the customer or prime enterprise before. Nor should they feel favored because of past or present personal relationships with customer or enterprise personnel. This favoritism is unethical and in many jurisdictions the subject of litigation.

  Also, subcontractors, like most businesses, are usually only as good as their performance on their last job. Sometimes the performance of a subcontractor with an outstanding past reputation falters. Many well-known names in many industries have for one of many reasons fallen behind their competitors. This is why the good Turnaround Lead will conduct their source selection, including rigorous proposal review, fact-finding, site visits, best and final, etc., as if it is the first time they are working with this subcontractor. The competent program save cannot afford to be held back by a subcontractor who no longer performs better than their competitors.
- Any subcontractor who attempts an "end run" around the prime to the customer to advance their own interests must be reprimanded. The subcontract wording should explicitly prohibit this and describe penalties that can be imposed. Usually the customer will appreciate the seriousness of this infraction and will support and respect a stern reprimand from the prime. In almost all cases, if this delinquent behavior continues, termination of the subcontract should be considered.

7.15 Chapter Highlights

- Face-to-face communication.
  - Up to 80% nonverbal.
  - Promotes fastest program pace.
  - High trust established.
- Virtual communication with care.
  - Does not replace face-to-face.
  - May be effective part of communication plan.
- The grace of colocation.
  - Provides immediate face-to-face communication.
  - All necessary team members or representatives should be present.
  - The "30-second" rule.
- If some program elements need to be geographically separated.
  - Maximum functional cohesion of tasks within each group.
  - Minimum communication coupling between groups.
  - Sometimes essential.
- Organization designs that directly map to customer's work breakdown.
  - Often close to ideal.
  - May reduce errors communicating program information with customer.
- Adhere to Program Plan and committed processes.
- Improvements to Program Plan by program process.
  - Often approved by a Change Control Board.
  - Decisions must be immediately communicated to team.
- Assign each task to one team member or leader.
  - Test: everyone on program should know who each lead is.
  - Team responsibility for task leadership is common contribution to program failure.
- Program meetings must be structured.
  - Meeting lead documents meeting purpose.
  - Invitation, agenda, attendance list, minutes, action items.
  - Results should be available to program team (within limits of sensitive data).
- KISS and the three levels of problem solutions.
  - Doesn't work.
  - Works.
  - Works and simple.
- The knock at the door—innovation!
  - Most breakthroughs come from individuals.
  - Leaders must give every suggestion due process.
  - Share good ideas with team.

- Team rhythm.
  - Periodic meetings.
  - Common processes.
  - Program standards.
- Find anomalies, errors, and problems soon.
  - Failure correction costs increase exponentially as product becomes more integrated.
  - Test coverage must be adjusted so failures observed do not reoccur.
- Watch the flank!
  - GANT may provide more visibility than PERT.
  - Listen to concerns from all levels and sources.
- Recruiting guidance.
  - Problem solvers versus problem workers.
  - Important questions.
  - Recruiting at large events.
- Subcontractors must be essential team members—nothing more or less.
  - Shrewd prime versus obedient prime.
  - No special treatment based on past affiliation.

Chapter Eight

Metrics—A Crystal Ball

Mentioning the word "metrics" to program leadership often causes mixed feelings. Some may believe it is a nice thing to do but that during the urgency of saving a failing program it brings little value. It might even be considered by some to be a hindrance.

I have witnessed large amounts of time being spent by program teams collecting and tracking measurements of what the program essentially already knew. They added precise, measured credibility to this knowledge.

But I have also worked on programs for which the analysis of the metrics was indispensable. They accurately predicted serious future issues months ahead of when they would have been observed during later stages of testing. In these instances, metrics were highly valuable—they were a "Crystal Ball."

8.1 A Little or a Lot

There is a misconception that maintaining metrics only applies to programs in which many copies of the same item are created. Examples of this would be large numbers of items created on an assembly line or a large number of responses to a questionnaire.

But even if this program is creating only one copy of a product or system, there are often many copies of some sub-element in this item. For example, a program may be performing a single study that requires accurately obtaining data with the same format from hundreds of different sources. Or a program may be developing one satellite that has a thousand identical solar cells on the solar arrays.

Therefore, most programs could receive great benefits from metrics if they are used correctly. Here are some guidelines for using this potentially valuable tool.

8.1.1 Measure Inputs, Not Just Results

I have observed this to be the most common mistake when using metrics (see Figure 8.1).

Let's take a simple example: Assume you are in the business of making and serving gourmet salads to the public. In fact, you have 30 stores selling these salads. Customers are waiting in line at the front doors. Business is great.

Being an astute businessperson, you maintain a weekly count of receipts of how many of each of the ten different salads you are selling. This is of course a metric of sales. For a while the count for each receipt is on the rise. Great news! But then slowly the sales per week for two of your favorites, "Salad Ambrosia" and "Salad Galore," start to fall.

Is this because customers are getting tired of these flavors? Is it because the cooks are making a mistake using the recipe? You taste these salads and find

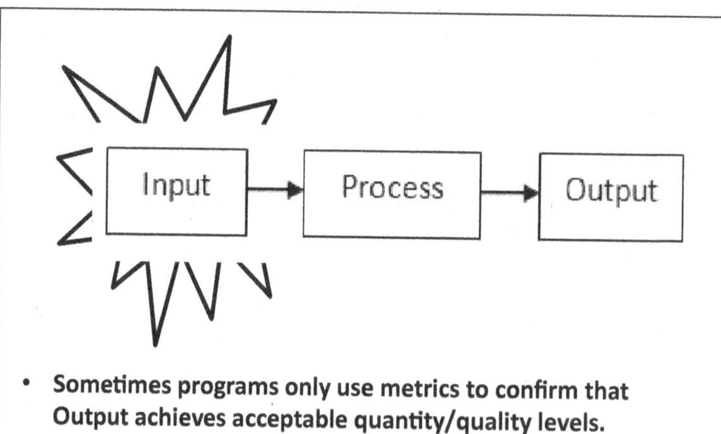

Figure 8.1 Unacceptable output values usually reveal a problem in a process too late for a turnaround program. Tracking inputs to a process with metrics will predict a future product failure.

they now have a bitter aftertaste. After further investigation, you determine that the radishes you're using are the source of the bitterness. With more detective work with the radish supplier, you discover they have gradually been using less water to irrigate these radishes to comply with drought restrictions, and this is why the radishes were becoming bitter. As quickly as you can, you evaluate five other candidate radish vendors and find one with radishes that taste ok. You immediately start using these in your salads.

But this radish issue caused serious damage to your business. It took time to find out your salads were tasting bad and to determine why this was happening. Then it took more time to solicit other sources for radishes, complete a trade analysis of their offerings, negotiate a contract with your new source, set up the supply line from your new vendor, and start distributing salads with the better radishes. During the time it took you to fix these salads, you lost a large number of customers. They not only have stopped buying Salad Ambrosia and Salad Galore, they're not buying as many of any salads from your business as in the past. They're even telling their friends that your salad business has "lost the formula." "They don't taste like they used to." As a result, you've lost 22% of your sales, and you may never earn all of it back. Is there some way you could have found this problem earlier and corrected it before it had this impact?

There is. Simply by keeping metrics on the *inputs* to your salad-making process, not just the output—in this case, by maintaining metrics on the acceptance testing of the radishes before they left the supplier, you likely would have been able to determine that their taste was unacceptable before you used them in your salads. You may have had to take these salads off the menu for a time while you found another radish supplier, but it's much better for a product to be unavailable for a short time, with the promise of it returning, than to sell a flawed version of the product.

But is there any way metrics could have been used to avoid incurring any downtime waiting to find an alternative radish supplier? Is there a way you could have maintained an uninterrupted supply of Salad Ambrosia and Salad Galore to your customers?

8.1.2 Monitor Input Trends and Not Just Acceptable Input Values

The answer is Yes! When monitoring the input items to your process, in addition to checking that the measured values were within some range of acceptance values, you could have also seen what the *trend* of these measurements were. This evaluation can be very powerful (see Figure 8.2).

Let's say that there is a battery of taste and chemical tests done on the radishes before they leave the vendor. A score is computed based on these measurements,

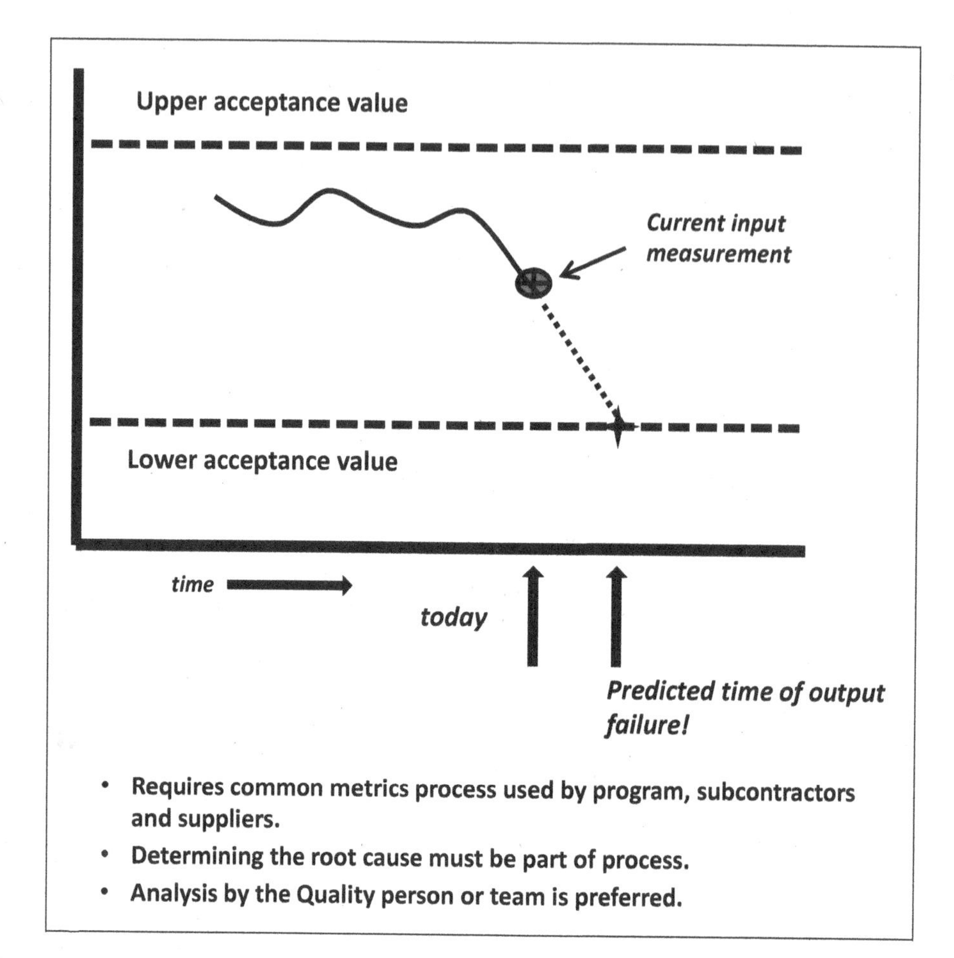

Figure 8.2 Metrics for inputs to a process must be checked not only for acceptable values but also for ominous trends. These trends can predict when the process output will fail, saving large amounts of program cost and schedule time.

and you judge that a value above 1.5 is acceptable. You noted that since working with this radish vendor, your scores have always been between 1.8 and 1.9. Great scores, you and your customers are happy. But then you notice one week that the score is 1.76. The following week it's 1.72. Sill the next week it's 1.68. You can extrapolate this trend and compute that in seven weeks you will be at or below a score of 1.5.

What did you get from this trend analysis? One, an early warning that one of the inputs to making your product was losing quality. And two, approximately when this input would be unacceptable and your final product would be flawed. It's your Crystal Ball!

Knowing when your product will be unacceptable will give you time to either correct the problem at your current radish vendor or evaluate and select a new vendor without stopping the manufacturing of your product. This can bring huge savings in profits and reputation to your enterprise—and save it from the disaster described above!

I have been on programs that have monitored product input metrics in the way described above very successfully. One such program predicted future failures in flight electronics units for a program three months before these units would have failed acceptance testing. In this case, tens of millions of dollars were saved, as well as the reputation of the program and the enterprise.

8.1.3 Determine Root Cause for Failing Trends

The importance of determining the root cause for a failure encountered during a program turnaround, and an approach for doing so, will be discussed in detail later. Suffice it to say, when a problem in the trend of product input metrics is observed, it should be analyzed with the same program failure review process that is used for any failure on the program. As for all program failures, the aim is to accurately determine the root cause so the failure can be corrected quickly and completely on the first attempt.

The mitigation of a bad trend in the process input metrics should be managed by the program's risk management process. This will allow the program to quantify the impact that would result from the troubled metric if it were not corrected, and then to implement corrective action in time.

8.2 Implementation

So, with this "magic" of predicting a major setback to the Program Turnaround with metrics, are there any downsides? The answer is that there are a few, but tracking metrics is usually still worth the investment in resources.

Monitoring just the outputs and not the inputs to your development process will be less work and may save money. But monitoring just the output metrics can simply confirm that your rate of acceptable output is per plan. It often applies as much to checking the program adherence to the production schedule as it does to tracking product quality. Monitoring the quality of the *inputs* to your program processes will add cost.

There are methods such as Six Sigma tracking that "qualify" a carefully fixed manufacturing process with a rigorous analysis and process demonstration. Less testing scope is performed on the output of this process, thereby saving total cost when many copies of something are built. However, this approach only has value if a much larger quantity of the item being tested is used in the product being created by the turnaround. Carefully examine a supplier's processes if they are proposing this approach for their work.

Most often the program's contract with the supplier, be they a vendor or subcontractor, will include the requirement to perform acceptance testing and report metrics on what they provide. This will usually provide the needed input metrics for the program's work. These are the process input metrics that should be monitored by the program for trends that predict failures.

The collection and analysis of metrics by the prime is usually done by the quality group if the prime is part of a big enterprise. For small turnaround programs, additional resources may be needed to do this work.

Any failure investigation of failed quality values or failing trends of a product provided by a vendor or subcontractor should be conducted by these suppliers. The fault determination process used by the supplier should mimic the failure review process used by the prime. This requirement should be stipulated in the contract for each of the suppliers. The Turnaround Program should receive a final report from the supplier that states the root cause of the failure or failure trend revealed in their product metrics and their correction action plan. The Program must approve the report before correction action is started.

Sometimes team members from the Turnaround Program will be asked to be members of the supplier's failure review team. An agreement with the supplier to provide this support must be established in writing in the supplier agreement contract or subcontract. The contents of this agreement will help the Turnaround Program determine the cost to the Program to furnish this support.

The Turnaround Program will have to decide if it is worth spending the added money for program/enterprise personnel to independently monitor quality metrics from the subcontractors and suppliers for their delivered products. This cost may be significantly lowered by requiring that the suppliers deliver their metrics in a common format at the same time. In many cases, it will be worth the investment to monitor this data.

8.3 Chapter Highlights

- Metrics beneficial when creating multiple copies of the same thing.
  - Multiple systems.
  - One system containing many copies of a sub-element.
- Measure trends in process inputs, even if they are at acceptable levels.
  - Facilitates correction before product fails.
  - High-value focus for the quality person or team.

- Root cause must be determined for failing input trends.
  - Prevents reoccurrence.
  - Establishes better understanding of product.
- Common metrics process and reporting formats should be stipulated in subcontractor and supplier agreements.
- Metrics downsides.
  - Added cost.
  - Resources to fix input trends that some say seem innocuous.
- Metrics upsides.
  - Major savings of program cost and reputation if employed correctly.
  - Avoidance of serious delays to program or project completion.
  - Enhances program team vigilance of product quality.

Chapter Nine

Contract Success

Managing subcontracts has been discussed earlier in this book, and I provide more details in this chapter. I also point out areas to pay extra attention to when negotiating subcontracts. This guidance applies to all Program contracts including the contract the Program has with the customer. Special emphasis is given to Program subcontracts.

9.1 Subcontract Management Organization

The following three roles of managing a subcontract are highly recommended. They cover all the interfaces and subjects needed to successfully select and execute a subcontract. If the subcontract is small, it will not require full-time support from the members of the management team. However, consolidating any of these roles into one individual will defeat the checks and balances needed to provide high-quality work.

9.1.1 Subcontract Manager

This person is the single point of responsibility for the selection of the subcontractor and management of the subcontract. They should report directly to the Turnaround Lead or their designee.

The other two leadership roles of Subcontract Administrator and Technical Lead (discussed below) report directly to the Subcontract Manager. The Subcontract Manager reports all contract status to the Program and is the single person responsible for the success of the subcontract. This person usually has a management counterpart who works for the subcontractor. The Subcontract Manager is the sole point of contact for this counterpart for conducting daily business.

The Subcontract Manager attends all the major subcontract meetings conducted at the various sites performing the work. They lead all the subcontract meetings the Program is responsible for conducting, and they record all action items for the subcontract and track them to closure.

The Subcontract Manager is responsible for estimating and managing the budget needed from the Program Turnaround to successfully manage and complete the subcontract. This includes the funds to pay the subcontractor as well the funds needed by the Program to manage the subcontract. The Subcontract Manager selects the other members of the subcontract management team and provides all needed subcontract team management services for the Program and the enterprise, including employee performance evaluations, team information exchanges, and subcontract cost and schedule tracking.

9.1.2 Subcontract Administrator

This person often receives special training and is certified by the enterprise to represent them in this role. They are empowered to sign all contracts and formal correspondence with the subcontractor. They have a complete knowledge of all the standard milestones needed to execute a subcontract, from soliciting bidders to closing the completed work. They coordinate source selection, coordinate negotiations, monitor that all meetings and design reviews have all the required work products ready and on time, make sure each meeting has the required attendance and is properly conducted and closed, and any other relevant actions pertaining to the administration of the subcontract. In addition, they work with the Subcontract Manager to make sure that all subcontract deliverables, reviews, and approvals are completed on time.

They work with the Subcontract Manager and Technical Lead to make sure all subcontract product deliveries are tested to satisfaction, provide the functionality and performances required in the subcontract and are ready to be officially accepted.

9.1.3 Technical Lead

The Technical Lead is an experienced expert about the product being delivered. As previously discussed, this product does not have to be a physical item. It can also be a study, an evaluation, a report, a design, a plan, a human resources service, etc.

The Technical Lead will usually have a history of academic or other special schooling about the product being delivered. In addition, they usually have a large amount of experience with developing this product and/or procuring this product from other sources.

The Technical Lead is responsible for checking all the technical content in all material provided by the subcontractor, including all proposals, contracts, contract deliverables, reports, technical discussions, and suggestions for improvements. This person may be asked to check that the work plans, work durations, and cost estimates proposed by the subcontractor are justifiable and reasonable. This lead attends all design reviews, product development reviews, and acceptance review/testing conducted with the subcontractor. They verify the technical compliance of all subcontract deliverables and give the Subcontract Manager and the Turnaround Lead their authorization to accept them. They, along with the Subcontract Manager and the Subcontract Administrator, must agree that the subcontractor has fulfilled all the requirements before closing the subcontract. They are involved in computing remedial action if the subcontractor does not provide the work described in the contract.

These three management roles provide a system of checks and balances that will ensure that the subcontract will be completed as negotiated and will be successful. For example, the Subcontract Manager would not be able to shortcut compliance checks of deliveries to recover schedule time. The Subcontract Administrator would object to this action because it would violate the testing process described in the subcontract. The Technical Expert would also object because inadequate testing would jeopardize the verification of the quality of the work delivered by the subcontractor. Each subcontractor reports to the Program staff and Turnaround Lead via their Subcontract Manager.

It is certainly possible for someone in any one of these three roles to support more than one subcontract at a time. In fact, it is not necessary that same three individuals form the management team for additional subcontracts if any of them is supporting more than one. Therefore, mandating that the three management roles described above must be assigned to each subcontract does not mean that a large and expensive full-time subcontract management staff is required for each subcontract.

9.2 Reviewing the Prime Contract and Subcontracts

Before signing a contract or contract revisions with the Program's customer or a subcontractor or other Program supplier, the words in this document must be reviewed with the utmost care. It is stunning how a simple but ambiguous sentence in a contract can cost the contractor and customer large sums of money and development time. To forestall any misunderstanding, all parties must evaluate each sentence in the draft contract as if it were to be interpreted in the most burdensome, risky, and expensive way possible. A real example of such a vague and dangerous statement for one program I was given was simply, "All subsystems shall be demonstrated in orbit." This wording should have been further discussed and clarified before the contract was signed.

9.2.1 An Ounce of Prevention

Again, these recommendations apply to a turnaround program or any program finalizing contracts both with their customer and their subcontractors.

Before signing, the contract must be reviewed extensively. This effort truly results in a "pound of cure." These reviews are performed at least by the Turnaround Lead, the Program's Contract Manager for the contract, the Contracts Administrator for the contract, the Program's Technical Lead for the contract, and the legal team hired or employed by the enterprise. Often reviews may also be of value to the Program by some of the Program's technical leads and task leads, particularly if their areas of responsibility are affected by the proposed contract or subcontract being reviewed. Also, consider having senior enterprise managers, technical specialists, and/or enterprise teams that are not a part of the Program perform an independent review of the draft contract.

The reader may get the sense that these reviews are a lot of work and may be overkill before signing a contract—but this intuition is not true. The investment of time and energy in thorough reviews usually pays back many times in reduced contract costs and management turmoil. In addition, it can avoid a memory of a bad business experience with a customer or subcontractor. Please see Figure 9.1 for a detailed review checklist.

When teams and individuals are reviewing the draft contract, they should check that all needed materials, facilities, and resources are addressed in the plan and cost estimate. They must make sure that the scope of personnel and materials needed to complete the expected work is covered. They must make sure that additional work needed to complete the contract is not hidden in proposed "contract options" or is assumed to be addressed in a follow-on contract or a separate contract between other parties. Every word in a proposed contract must be examined to determine if its meaning is ambiguous or dangerous to the Program. This includes the words in the "standard" Terms and Conditions section as well as other so-called contract boilerplates.

The parties negotiating a contract will tend to interpret the wording in a way that is beneficial to them—this is just human nature. By not accepting

Summarize what services are being contracted.

What type of contract is this?

Service
Goods - Product

Document the contract decisions:

Funding Availability.
Cost/Benefit Analysis.
Available Program or Customer Resources.
Legal Constraints to Contracting?
Contracting with Current or Former Enterprise Employees?
Method of compensation and billing?
Evaluate coverage provided by any existing and anticipated audits.

Risk Assessment and Monitoring Plan:

Were customer risks assessed prior to entering into a contract?
Have the risks been assessed to execute the contract?
Does the risk assessment form the basis of the monitoring plan?

Contract Provisions:

Is the scope of work clearly written and defined?
Are performance measures required, and do they satisfy needed project outcomes?
Are hold harmless and indemnification provisions included?
Are liability and industrial insurance provisions adequate?
Do contract termination provisions protect the rights of the contractor?
Was an appropriate compensation method selected and identified in the contract?
Is coordination with other project team members an issue?

Technical Assistance:

Will this contractor need technical assistance outside of the enterprise?
How will technical assistance be provided?

Progress Monitoring:

Is there a monitoring plan in place? What monitoring activities are listed in your plan?
If corrective action is needed, is there a corrective action process required?

Post-Contract Follow-up:

Any activities needed?
Is follow-up on audit findings needed?
Must Program objectives and outcomes be evaluated, assessed, and approved?

Figure 9.1 Each contract in the Turnaround Program, including those with the customer, subcontractors, suppliers, and service providers, must be reviewed in accordance with a standard checklist.

ambiguous wording, you can often avoid unnecessary and permanent damage to a business relationship with the contractor. Not making the investment to scrutinize every word before signing the contract is unfortunately a common mistake in both new programs and turnarounds.

Often it is necessary to "take too long" to thoroughly review a proposed contract with a customer, subcontractor, or supplier before signing it. All parties may be uncomfortable with the length of time it takes the Program to complete their review. But again, assume a role of devil's advocate against the Program when interpreting each sentence. Make sure any ambiguity is clarified. Have an appropriate, qualified legal team review the contract for clarity and for wording that might be detrimental to the Program before signing.

9.2.2 Going Native

When a program team member is assigned to work closely with a subcontractor and possibly even reside at their facility, this person will naturally understand and even share the rationale for the successes and failures the subcontractor is experiencing. The Program representative may even begin to feel like a member of the subcontractor's team.

With time, this empathy with their assigned subcontractor can lead to their arguing less for the interests of the Program and more for the interests of the subcontractor. This is often referred to as "going native." It is common and very human—it takes extraordinary discipline and experience on the part of the Program representative to prevent this from occurring.

In its extreme, a program representative going native can be confusing and embarrassing to the Program. A Turnaround Program team member who represents a subcontractor and who is able to objectively describe both viewpoints in a dispute can provide important information to the Program. Indeed, this is one of the tasks you have hired them to perform. But, if they are suddenly blaming the Program for all the problems the subcontractor is having or they are staying conspicuously silent about issues with the subcontractor that are adversely affecting the program, then it is time to review their assignment.

Again, going native is a natural human response. The extent of it differs among individuals but it occurs quite frequently. A reprimand or psychotherapy is both unwarranted and ineffective!

The best solution is to periodically rotate program personnel through this role, particularly for onsite representation. Rotating representatives brings a constantly renewed and objective perspective to the interface between the Program and the subcontractor and facilitates fresh ideas that can benefit both parties.

9.2.3 Internal Reviews

As mentioned earlier, having an enterprise team that is independent from the Turnaround Program team conduct a review of a draft contract is very beneficial (see Figure 9.2). Members of this team are unbiased by schedule pressures, recent correspondence with the subcontractor or customer, long affiliations with some proposed solutions, etc. Their review may not only improve the content of the proposed contract but may offer new approaches that should be considered.

- **Enlist knowledgeable professionals who did not write the contract.**
- **Contrive most disadvantaged interpretation that each contract statement may have,**
  - **Example: "All safety systems shall be demonstrated in final product."**
  - **Independent review will help isolate ambiguous and potentially dangerous wording.**
- **Never assume wording will be interpreted in favor of the program or project.**
- **If acceptable, request a customer review.**

Figure 9.2 Incomplete reviews of program or project contracts is a common cause of unplanned expenses, delays, and even program failure. These guidelines help with scrutinizing every sentence.

9.2.4 Customer Reviews

In most cases, the customer should be given the opportunity to review the draft subcontracts and other supplier contracts. There may be rare instances in which proprietary data from the subcontractor prevents this. The customer may respond that it is not necessary for them to conduct a review, but they should be given the opportunity.

Inviting the customer to review draft subcontracts has many benefits. It provides another independent review by a qualified member of the team, and it gives the customer insight into the Turnaround Program's subcontractor evaluation process. This gives the Program the opportunity to ratify how thorough and disciplined their selection/negation process is. Also, if the customer reviews and endorses the subcontract, they implicitly share in the success of the contract work.

If the customer has a special preference for a particular subcontractor, vendor, or team member, it is important to give high credence to it. Most often, a customer-recommended source has successfully worked with the customer before. This source may have some special capabilities that are not obvious and make them a better source. However, if the costumer-preferred source appears to jeopardize the Program, it is appropriate and necessary for the Program to inform the customer of their reason for concern.

9.3 Chapter Highlights

- The basic management roles.
  - Subcontract Manager.
  - Subcontract Administrator.
  - Technical Lead.
- Role definitions provide checks and balances.
- Each role assigned to a different person to prevent conflict of interest.
- Roles may be performed part-time.
- Carefully review contract before signing.
  - Internal.
  - Customer.
  - Legal.
- Beware of management "going native."
  - Long-time/on-site affiliation can bias judgment.
  - Consider mandatory periodic leadership turnover.

Chapter Ten

Expectations

The new Turnaround Lead and the program leadership team must promptly establish the updated work expectations as the Turnaround is started. It must be conveyed by the Turnaround Plan, by their words and example.

In some cases, these expectations may already have been a part of Program culture, but the request for team commitment from the Program was not clear.

10.1 Laser Focus on Results

A firm expectation must be established with each member of the Program that a fundamental part of their job is to commit to a completion time for the work they are doing and to achieve that commitment. It should be emphasized that their predicted completion dates are not a "goal" or a "hope"—they are a commitment. The estimated completion time must not assume that progress will be made without encountering development anomalies or other incidents. Their estimate should include margin for unanticipated problems. However, each team member must know they are completely responsible for achieving their commitment on time even if they use all their schedule margin. Demonstrating this performance is part of the evaluation of their work.

When program members make a schedule commitment, they should provide a completion time that they know with high confidence they *will* achieve, not one that they *hope* they can achieve. Good program leadership must be aware that some team members may want to offer completion dates that are excessively conservative with huge margins to virtually eliminate schedule risk.

My experience is that this approach quickly becomes obvious to leadership and is best mitigated with a frank discussion with the team member to clarify their approach to scheduling.

If unforeseen delays occur during the completion of their work task, team members must know they are still obligated to meet their commitment dates. As a result, there may be times when the team member must work extra hours and make personal sacrifices to meet their deadline.

The emphasis on schedule commitment must be a fundamental working habit for all members of the Turnaround Team. As mentioned earlier, the measure of the contribution a team member has accomplished must be the results they have achieved, not just the time they've spent getting to the results.

Many times, delivering the planned results on time requires devoting most or all of a team member's attention to that task. Often, the responsibilities of a team member on a turnaround are too complicated to allow them to work 100% on one task. But leadership should encourage the team to minimize or eliminate multitasking.

Many times, the hardest part of any task is the last 10% to 20% of the required work. Often the best way to get a task done is to work exclusively on it near its completion.

It will be important for the Turnaround Program leadership to keep program planning and schedules current and easily available for review by all members of the program team. This status information should be available to the team members online if possible. This will provide a clear and current

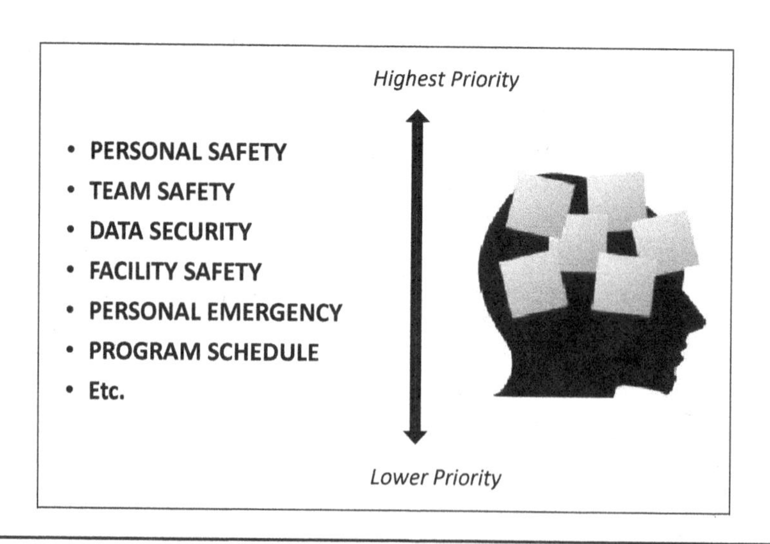

Figure 10.1 Providing each team member with a hierarchy of workplace priorities may seem a bit tutorial, but it will greatly reduce confusion in the event of a workplace crisis. This list is an example.

perspective on the importance of each of their accomplishments toward achieving the Turnaround Commitment. The status of near-term planning and schedules should be covered at the morning status reviews recommended earlier in this book. The perception of the high importance of the work performed by each team member will be increased by emphasizing how critical its completion is toward achieving the Turnaround Commitment.

An aid to help the team members to direct their energies toward completing the most important tasks is to review what the team's most basic priorities should be. A popular example of this hierarchy of priorities is shown in Figure 10.1. This simple hierarchy should be provided in writing and will be a great help for the team members to quickly and confidently resolve high-priority distractions while completing their assigned work.

10.2 The Long Hours

Achieving schedule deadlines is critical when executing a Turnaround Program. Because unforeseen problems will be encountered that may consume all schedule margins, the Turnaround Team must be prepared to work extra hours to maintain the schedule plan. Following are a few actions leadership can take to motivate this.

10.2.1 Set the Example

Leadership must "walk the walk." They must visibly take on specific program tasks and achieve committed milestones despite the setbacks they may encounter.

Program leaders should be on site before the first team member arrives and after the last one leaves. Exceptions to this would include personal business or travel. In this event their representative should be present. If the work is being performed in 24-hour shifts, the leaders or their designees should be accessible around the clock, by phone if not face-to-face.

Program leaders must be prepared to demonstrate that they are making the same types of personal sacrifices they are asking of their teams. These sacrifices may include disruptions and postponements of time they had put aside for recreation, hobbies, vacations, and more.

10.2.2 Reminder of Importance

The Turnaround Team should be reminded that they are special by virtue of the fact that they were chosen to be members of the Turnaround effort and by the high importance of the program they are putting back on track. They should understand that the high expectations placed on their performance are exceptional.

10.3 On Call

Team leadership must have current contact information for all team members. Sometimes team members must be on call around the clock. Special care must be taken by leadership to never abuse this privilege. If a leader calls someone at 2:30 AM, it must be because his or her participation is essential for the continuing success of the program.

When off-hours calls are made, they must be accompanied with sincere apologies and gratitude for the extra effort the team member is making for having their personal life interrupted. A good team member will understand the need for these intrusions and take pride in responding to them professionally—that is, if the call is necessary!

10.4 Personal Sacrifices

Personal sacrifices go beyond working longer than standard hours. These can include postponing weekend plans or vacations, being on call around the clock, working nonstandard shifts, etc. Leadership must acknowledge that these sacrifices will affect the lives not only of the team member but also of their families. The Program must compensate them for any fees or penalties they may incur to support the turnaround work, including cancellation of hotel, a tour, or airline reservations; increased costs caused by revising vacation times; purchases that are not used because of a required schedule change; etc.

10.5 Keep Raising the Bar but Have Their Backs

Please refer to Figure 10.2.

10.5.1 Again, By Example

The Turnaround leaders must set the example of continually challenging themselves to meet higher standards. As we perform a task, we get better at it. Work that may have been daunting at first becomes easier to complete as we find ways of being more efficient. This is the time to add a few more elements to our load as a leader. Visibly take on additional challenges. No team member should be

Expectations

- **A focus on milestone completion on time.**
- **Individuals acknowledge task ownership.**
- **Work one task at a time.**
- **Devote the effort needed.**
- **May be on call.**
- **Personal time may be affected.**
- **Leaders will set uncomfortable targets.**

Defense

- **Protect against:**
  - **Malicious criticism**
  - **Potential force reduction**
  - **Time needed for personal issues**
- **Sell the attributes of reports at all opportunities.**
- **Forgive for mistakes made in earnest.**
- **Document basic work priorities.**
- **Current program completion status available to whole team.**

Figure 10.2 Setting challenging expectations for the team while vigorously defending the team members may seem contradictory. However, it galvanizes team dedication and motivates special efforts to innovate.

completely comfortable when they are working to save a program. If they are, they are probably not challenging themselves enough.

The Turnaround leadership should provide visible evidence that, for example, they are taking on additional tasks or adding metrics to their weekly status reports. They must set the example that they are learning to do their current job faster and more efficiently and are taking on more work to keep their level of contribution to the Turnaround as high as they can manage.

10.5.2 Grow the Demands

The sensitive and observant program task leader should gradually increase the task load of their reports as their reports learn to complete their current tasks faster and with fewer schedule risks. Sometime this is initiated by asking team members to accomplish a little more than they estimated they could do.

The size of this request will depend on the individual circumstances. For example, some team members may estimate their completion dates very optimistically, while others will build a lot of margin into their estimates. Some team members may be comfortable achieving the milestones for the work currently assigned to them, while others are still struggling to complete their assignments on time.

10.5.3 Have Their Backs!

In my experience, nothing garners dedication by team members to leadership and the enterprise more than knowing that their leadership is protecting them and has their best interests at heart in the decisions they make. I have witnessed and have worked for program leaders who are sometimes blunt or even gruff but are 100% ready to stand up for you if you are being unfairly criticized. These special leaders will pick you up if you fall and give you a second chance, often with a little coaching. These special leaders can differentiate which team members are trying with all their hearts to do the best they can and which are only trying to convey the impression of hard work.

The good leader will protect their team members from undue criticism from both inside and outside of the program.

Unfortunately, sometimes a team member can become unreasonably critical of one or more other team members. This criticism can go beyond the usual banter that teams might develop when under pressure, possibly extending to hurtful sarcasm or downright abuse. If their victims confront them, the situation can escalate to a point where intervention by leadership is required to prevent a severe demoralization of the team.

I cannot over-emphasize how deeply hurtful this behavior can be, both for the victim and the team as a whole. This antagonistic individual must be confronted promptly by leadership and told to immediately stop their behavior. Leadership must be specific about what behaviors are unacceptable. This is not the time to evaluate the merits of the often obnoxious comments made by this individual. If the behavior continues, the perpetrator must be removed from the program.

By taking action, leadership demonstrates to the program team not only that they hold all team members to high work standards but also that they will act to protect their teams from criticism and abuse.

Criticism from outside the Program can also be hurtful and destructive to the team. All critiques and improvement suggestions brought to the Program from outside it must be fairly evaluated for legitimacy and value. If they do, the

Program should respond with a thank you and determine the value, cost, and urgency of implementing the correction. If they do not have legitimacy or value, leadership must inform the outside source that this is so. What is important is that the teams know their leadership takes any outside criticism seriously, examines it fairly and clearly, and refutes it if it is wrong. It is up to program leadership to stand up to unwarranted criticism from outside the program.

Program leadership should always sell the attributes of their individual team members to enterprise management. Whenever possible, leadership should be active in getting their team members promoted when this is deserved.

Of course, some team members are satisfied working on their current assignments and keeping the roles they have. These types of team members are part of any well-balanced team. But for those struggling to grow their responsibilities, leadership should demonstrate that this is a priority for them too.

As an example of another way leadership shows that they will stand up and protect their teams, many years ago, I was an engineer on a program of about 2,000 people. Like many of my peers, I was struggling to find a way to make the biggest contribution I could to this effort. Yet with just a two-week notice, this program was cancelled.

At the time, we were part of an enterprise of over 25,000 people. Often in these circumstances an enterprise will respond by finding positions for employees they have determined to have very high potential, then terminate the rest. They may call this making the "hard decisions." Some executives even get bonuses for this response.

Unfortunately, this kind of response may improve the quarterly numbers but damage the long-term health of the enterprise. Most of those employees staying after the cutback will feel they are standing on a trap door. The team members' first priority now becomes self-preservation, not innovation and doing what the program needs the most. This is bad for a program turnaround. Sometimes the performances of the remaining high-potential individuals end up being disappointing.

The response to the cancellation of the program I was on was an announcement by the president of the enterprise that everyone would still have a job. It was mandated that the displaced team members be absorbed by other programs in the enterprise. And that is exactly what happened. Some individuals went directly to a new role, while others improved their skills for a short time, on enterprise overhead, while waiting for their next assignment. Many team members actually found a better professional growth opportunity after they were assigned to a new program.

Most important, a passionate and permanent dedication to the good health of the enterprise was developed in those employees who had been orphaned by the program cancellation; they continued to give their very best work out of gratitude that the leadership and the enterprise were looking after them, and they continue to do so decades later. Unquestionably, the improvements to this enterprise's business and profits derived from these enthusiastic and dedicated team members was far greater than any short-term financial cost of retaining them.

It is important for leadership to envision the future role of every one of their reports. In doing so, often the leader will see a capability in a team member that he or she does not see. This can be a surprise to the team members, as they may never have imagined themselves in the role their leader is suggesting.

If an employee wants to develop their potential talents more and if their leader carefully tailors their job assignments to support this direction, this growth opportunity can be immensely valuable to both the employee and the enterprise.

Most team members will respond to these opportunities with enhanced dedication to their leaders and to the enterprise. Aggressively trying to help each team member be the best they can be demonstrates a high level of leadership. Even if the leader's vision of the team member's promoted role is not achieved, this exercise teaches everyone involved something more about themselves. Plus, if done fairly, it can be a powerful demonstration to the program and enterprise team members of program leadership looking out for their best interests.

10.6 "You Can Pull the Line"

In one Japanese car factory, leadership installed a long rope-like line above the entire length of the assembly line. If anyone reached up and pulled the line, the entire assembly line would stop moving. The assembly line workers were told that if they saw a quality defect or any other problem in the assembly procedure, they could reach up and pull the line; the assembly would halt and would not resume until the issue was resolved. This process was a potent step toward bringing the whole manufacturing team together to ensure high quality in every car built.

I tell the team members on any program we are saving that they can "pull the line" for any reason—not only to maintain the quality of the work the team is creating but also for the health and safety of the team member and their families. Some readers may wince at the thought of this, but after leading many programs, I have found this option to be seldom invoked and never abused.

The "line" has been pulled for vendor parts suddenly failing acceptance testing, for mistakes caused by team exhaustion from long work hours, and for new issues that could reduce team safety. In each case, the issue was resolved completely before work on the program continued.

A few times an individual had to pull the line for a personal problem. I told the team members I did not have to know the details of their circumstances, because I had faith that if a team member pulled the line they had a good reason. It could be for individual family needs, heath issues, or just the need to let their mind rest due to exhaustion—anything. I asked them just to let me know they were leaving and to provide an estimate of when they would return. We had someone temporarily replace them or we re-sequenced the program tasks to accommodate their absence.

While hundreds of people have worked for me for years on large programs during high-pressure turnarounds, I am pleased to say that fewer than five pulled the line for personal reasons. More importantly, team members seemed to be more successful at keeping their work stress in check by knowing they had this option.

Having the option to pull the line is further evidence to the team that their health and well-being is first priority for program leadership.

10.7 Chapter Highlights

- "Laser" leadership focus on completing program milestones.
  - Individual/team *commitments*, not just goals.
  - Closure must be complete.
  - Leadership shows appreciation for milestones completed.
- Daily work completion status accessible to all team.
- Encourage completing one task at time, accomplishes more than multitasking.
- Review basic hierarchy of priorities in work place.
  - Reduces confusion in dynamic turnaround environment.
  - Should be written.
- Devote the work hours needed.
  - Achieving committed milestones is paramount.
  - Leaders set the example.
- Leaders set the bar high but have team's back.
  - Set bar often higher than comfortable.
  - Shield team against worthless criticism, reductions, etc.
  - Leaders sell their vision of each of their reports to the enterprise.
- Establish that anyone may "pull the line."

Chapter Eleven

Ethics Are Essential

A high level of ethics is essential for a program to be on step and for a turnaround program to be successful. However, many programs, including turnaround programs, do not establish this. Some program leaders may feel there is not enough time to make an investment in training good ethics during the urgent schedules of a turnaround. Some leaders might even consider that turning a troubled program around is an emergency action that might excuse occasionally bending ethical rules a little to be successful.

In fact, the opposite is true. Good ethics, exercised by all Turnaround leadership and team members, provides an extra layer of trust and mutual respect in the program team. This makes the Program Plan run more efficiently and be more successful.

What are good ethics? They include complete truth and honesty among all team members, plus an equally high respect for the contributions from everyone, regardless of background. Fellow workers are trusted to provide the most accurate status and estimates they can. Good ethics assumes there is no gaming or political posturing among groups working in the Program. The team receives all suggestions, criticisms, and work products in the same open manner regardless of the sex, age, ethnic background, sexual orientation, seniority, etc., of the originator. Personal items brought to the work area by the team members never "disappear." There are no prejudicial, insulting or threatening expressions in any form made by any Program team member to others. All team members have a high level of trust and respect for each other.

However, this high level of ethics is not created just by good leadership examples—it also requires the following.

11.1 Regular Ethics Meetings and Distribution of Written Reminders

This may seem like a parochial way of reinforcing high levels of ethics. But, it sends a strong message that the practice of good ethics is a very serious part of working on the Program.

The general principles and scope of ethics should be reviewed periodically during regular team meetings. It is also a good investment to conduct periodic special meetings to specifically train and discuss ethics. It can be very helpful for the meeting members to split up into subgroups and role-play ethics infractions, while team members and leadership determine the appropriate responses.

11.2 Equal and Swift Due Process

Any ethics violation during the Turnaround must be adjudicated swiftly and with the same fact-supported due process that would be applied to any infraction. As a result, word will get out to the program team and beyond that breaches of ethics are taken seriously and treated fairly. If the enterprise is big enough to have a separate Human Resources group, this process should be conducted by them.

11.3 Leadership by Example

As with all policies for the Turnaround, leadership must be unwavering examples of good ethics. Each leader must follow one set of ethics rules in any formal or informal setting, whether on the job or after hours. Their behavior must provide little room for levity. The subject of ethics must always be treated by leadership with the highest importance.

Good program ethics fosters a special bond among the team members. They have made a special investment in an effort to cement close trust and support of each other. As a result, they will develop an even higher level of pride in working with each other. Applying strong ethical behavior in their work can even become an unwritten requirement for team members to stay in what some may consider an elite program turnaround team.

As mentioned earlier, high team ethics actually increases the speed with which the Program fulfills the Turnaround Commitment and better assures that high-quality work is delivered. Team members will communicate with a high level of trust, knowing that good ethics discourages hidden agendas.

Trusted communication results in faster and more accurate communication among team members, which facilitates a higher program development pace.

Also, high ethics assures that all team members, regardless of background, have an equal opportunity to be heard and to contribute to the program save. This is necessary for a successful turnaround. Only by evaluating every good idea fairly and on its merits can the most beneficial innovations and fastest progress be achieved while fighting to save a program.

11.4 Chapter Highlights

- Ethics fosters added pride and self-respect in the program.
  - Regular ethics meetings with written material.
  - Adjudicate infractions "by the book."
  - Leadership must demonstrate clear example of good ethics.
- Enhances trust among team members, faster completion of tasks.
- Insures all team members may contribute equally.

Chapter Twelve

Effective Leadership and Basic Planning

In this chapter, we first review some important issues presented earlier in this book. Then I delve into more detail about how the initial turnaround task plans and schedules are established and integrated. I highlight the roles of professional schedulers and the leadership team to quickly assemble and validate the Program Master Schedule. Completing this work quickly and completely is necessary to successfully start a Turnaround Program.

12.1 Review and Elaboration

As discussed earlier in this book, one of the first tasks to perform when taking on a role as a Turnaround Lead is to divide the needed work to achieve the Turnaround Commitment into its component tasks. This task breakdown must then be used to define the work teams needed in the new organization. An important criterion for defining these teams is to make them as independent from each other as possible. As discussed earlier, this high independence is established by organizing the teams such that each contains task functions that are highly related and interdependent (high functional cohesion) while requiring interfaces to the other task teams that need as little information as possible (low coupling among teams).

This will allow the task work to progress in each task team with minimal impact on the work of the other teams. It also provides a map for dividing the tasks such that each can achieve some degree of parallel development, thereby reducing the time to achieve the Turnaround Commitment. The processes for achieving maximum cohesion and minimum data coupling is discussed earlier in this book.

The customer will often provide a work breakdown structure (WBS) or similar product when describing the work they want done. Often this can be used as the map for identifying the independent program work tasks. This customer-provided structure may not separate the work tasks per the strict rules of coupling and cohesion, but it has the advantage of being a division of work the customer is familiar with. And upon analysis, this work division is usually close to ideal. It should be left to the judgment of the Turnaround Lead to decide if the program should use the division provided by the customer or suggest changes to it. The Lead should confer with the customer before making this decision.

After the division of tasks has been established, the Turnaround Lead must assign one responsible lead for each of them. A potential source for a task lead assignment is a team member with in-depth experience with the task but who may not have had a key leadership role before. This concept was introduced in Chapter 5.

In the heat of an urgent turnaround, the leader of each task team must fully understand the substance of the task. Regardless of whether the task is designing and building a product, establishing a new process, performing a trade, or providing a service, the leader of that task should have expert knowledge in the area to support the fast pace of a program save.

Look carefully into the membership of each newly defined team. As discussed in Chapter 5, the best leader may not be the individual with the most formal management experience or training, but the team member with an expert understanding of the task the team is working on. They usually have a complete perspective on the work to be done and a realistic understanding of the time it will take to complete this work. They may not have had an official leadership role in the past, and this may be the time to give them this assignment. These kinds of team members usually know how best to get the program tasks completed and have a realistic understanding of all work needed, the likely issues that will be encountered, and the time to complete it.

From my experience, it is surprising how well such new leaders accept this responsibility and successfully lead their assigned tasks to closure. While maybe a little clumsy or self-conscious at first, they are very motivated to get the job done with high quality using all the resources of the team. In return, the task team respects this lead for their highly competent, result-focused direction, and they follow him or her with a high degree of trust. They respect this lead for their expert knowledge and for the quality work they have completed in the past. An assignment of task leadership from this source usually turns out to be a great benefit both to the program and to the Turnaround Lead who has given them this leadership role.

Making leadership assignments in this way has the primary goal of assuring the success of the new Program Plan. However, often it is also an opportunity for the enterprise to identify future leadership candidates who may not have been obvious in the past. Enterprises highly value pragmatic leaders who are fair, have the passion for finding the shortest path to getting their assigned tasks completed, and then successfully accomplish the resulting plan with a team.

Once the tasks and task leads have been selected, the next step is to construct a schedule for each task team for the work needed to be completed. This is best accomplished with the assistance of one or more professional schedulers trained and experienced in recording important milestones, verifying work durations they are provided, and establishing the dependencies among them. They will maintain an objective perspective of the developing plan and can create a draft schedule quickly.

When working with a scheduler, each task lead should be given the freedom to articulate the top-level series of steps they believe they need to get their task completed. If they attempt to assemble a detailed schedule at this starting point, they may have to reiterate it many times, and the results may not yet have uniform planning detail. Because of the schedulers unbiased view of all the work steps needed to complete the task, they will identify portions of the task sequence that appear to be lacking identified interdependencies and work details. They will work with the new task lead to fill in these planning areas.

Working with a scheduler, the task team lead can often communicate their top-level work plan and schedule in as little as 30 minutes. During this time, task lead and the scheduler must constantly check that their proposed task schedule is consistent with the guidelines and schedule constraints of the Turnaround Plan. With an adequate number of schedulers (one scheduler per five to ten task teams), the process of creating the top-level schedules for all task teams in the Turnaround will usually take no more than a day.

After completing the top-level plan and schedule for each task team, the scheduler will then integrate these results into a Program Master Schedule to achieve the Turnaround Commitment. This new Master Schedule must be compared with the customer's master schedule and any special schedule constraints to verify compatibility. Then, the Program's subcontract managers and those managing any additional suppliers must have their schedules integrated into the Master Schedule to verify compatibility.

By assembling the draft Master Schedule, any incompatibilities of estimated completion times between the customer schedules, the task team schedules, and the subcontract/supplier schedules will become apparent. Usually these incompatibilities are minor, but if serious, resolving them must receive high program leadership priority. It is essential that all team schedule incompatibilities be known and solved before embarking on a high-paced Turnaround Plan. Please note that at this critical planning time, good schedulers may also identify incomplete task interdependencies and other subtle schedule deficiencies in the new Master Schedule that must be solved.

This will result in the first time a well-substantiated completion date for the Turnaround Commitment is derived. If this date does not support the customer's needs, the Turnaround Plan may have to be re-worked or have work scope removed, or, as a last resort, the Turnaround leadership may have to conclude that the Program cannot achieve the required completion date.

After the Program leads, Subcontract Managers, supplier managers, and schedulers have completed a Master Schedule that they believe is complete and acceptable, they should present it to the Turnaround Lead for final review and approval. The Lead will usually present the new Master Schedule to the customer and enterprise management for their approval at this time. These are very important reviews. If the schedule is approved, it will become the Master Schedule for the Turnaround and will be used to track the daily progress of the program. In the future, changes to the plan will only be allowed by a formal change control process, as discussed earlier in this book. This Master Schedule documents the individual milestone commitments from all Program leadership, and the Turnaround Lead. Concerns and comments from individual team members about the schedule while the program is underway will be evaluated, but at this point, Program leadership has assumed the responsibility to complete their work as scheduled.

I must reiterate that by signing up to the program schedules, leadership is signing up to a promise. As described in earlier chapters if some work takes longer than scheduled plus reserves, the task teams are expected to devote the extra effort necessary to stay on schedule. If a delay in a task schedule is caused by a problem so large it will be impossible to comply with the original schedule, then the Master Schedule must be revised via the formal program schedule change process that is described in detail in the Turnaround Plan.

12.2 Chapter Highlights

- The work needed to achieve the Turnaround Commitment must be decomposed into its most independent pieces via process or customer's work breakdown structure.
  - Reduces planning and coordination errors.
  - Identifies portions of tasks that may be executed in parallel.
- Assign most knowledgeable person to lead completion of tasks.

  - Should have reputation for being highly informed and focused on closure.
  - May not have much management experience or formal training.
  - May not be the original task lead.
- Selecting effective non-traditional leaders may be a lesson to the team.
  - Demonstrates that knowledge of the work domain is important.
  - Demonstrates that focus on work quality and completion is important.
  - Demonstrates leadership's willingness to advance team members to new responsibilities.
- Each lead meets with a scheduler and lists the steps and durations to complete their task.
  - First draft in less than one day.
  - Turnaround Lead must approve.
- Schedulers integrate task schedules into a Master Schedule.
  - Integrate with customer schedule, subcontractor/supplier schedules, and other constraints.
  - Reviewed and approved by Turnaround Lead, Program Leadership, customer, and enterprise management.
  - Post Master Schedule for the Turnaround Team.

Chapter Thirteen

Motivate Continuous Improvement

In this book I frequently used the terms "innovation" and "continuous improvement." My meaning of continuous improvement is that the search for innovation should be a daily activity by the program team members when executing the Program Plan.

Earlier in this book, I discussed how often excellent innovations come from a single mind. They usually originate as an inspiration and can come to us at any time. I discussed how, when a team member approaches me and says, "I know this sounds like a crazy idea, but . . . ," the suggestion that follows can sometimes be profoundly valuable—even a breakthrough—to the Turnaround Program.

The challenge to the leadership of a turnaround is that these improvement suggestions cannot be forced to occur; however, there are ways leadership can encourage these inventive ideas, allow them to be easily brought forward, and evaluate them with an engaging process. Program leadership must encourage the Turnaround Team to bring forward and promote their innovations and improvement ideas. An open door to all design and processes improvement ideas from everyone on the Turnaround Team is necessary to save a troubled program.

Keep in mind that these brilliant solutions are not only provided at the doorway to your office. They can be brought to Program leadership during a meeting, in the hallway, in the parking lot, anywhere. I have received phone calls with excellent new ideas while at my home.

So how can leadership encourage continuous improvement and keep the door wide open?

13.1 Program Leadership Asks the Program Subject Matter Specialists "What If?"

This requires that the leader have a general understanding of the subject matter the specialist represents, but it's okay to confess that you're an amateur and your questions may be naïve.

The "what if" question should have real merit—don't suggest an idea you know has no merit just to get people thinking. Examples could be, "What if we get feedback from graduates who've been in the workforce for a year as to what they wished they were taught more about in our curriculum?" Or, "What if we use radio signals from other satellites to maintain spacecraft attitude knowledge if the onboard sensors fail?"

Many times your improvement suggestion may be a little outside the field of view of a specialist who has concentrated more on solutions that have a high likelihood of working. However, your questions may assure them that it's okay to seriously consider options that are unconventional and may be risky. This provides an example to the specialist and the rest of the Program team of a team member (in this case a leader) struggling to find a way to improve performance, reduce cost, increase reliability, reduce schedule time, etc.

As a leader, your example of an attempt to innovate with questions and new ideas emphasizes the importance for all team members to constantly be thinking of ways to improve the product. There may even be an occasion when that naïve "what if?" question has merit and is applied!

13.2 Share Good Suggestions, Even if They Fail

This point has been made before in this book. Innovation is sometimes like mining for gold—it may take a lot of digging to find a nugget. During a program turnaround, the team should be encouraged to dig constantly.

At Program meetings and all-hands meetings, share some of the improvement ideas from the team, even if it turned out that these ideas were not implemented or they were implemented but failed. As a leader, you're celebrating the sincere attempts by fellow team members to improve the Program. You are clearly placing a high value on these attempts, which should encourage the whole Program to keep digging.

You can remind the team that it took Thomas Edison hundreds of tries to find a material for the filament of an electric light bulb that would not immediately burn out. His eventual discovery is still in use in incandescent bulbs manufactured today.

13.3 Recognize New Ideas That Have Improved the Program

When a suggested innovation succeeds in improving the Program, its value must be announced to the whole Program and the enterprise (unless, of course, the originator does not want their contribution shared with others. This does happen.) Most team members will be proud to be known as having contributed an idea that helped the Turnaround to be successful.

By explicitly describing the improvements to the Program brought about by each successfully implemented innovation, the high value of continuous improvement is clearly demonstrated to the team. In addition, it reminds the team members that developing improvements for the Program is important work and is considered favorably when their work performance is evaluated.

13.4 Never Punish for an Idea That Does Not Work

As we know, a punishment does not have to be an explicit act. Contributors can be very sensitive to the response of their leadership to their work performance; even a vague expression of disappointment to the originator of an improvement idea that did not work can be damaging. Innovations and other improvement ideas will cease to appear if leadership shows this kind of disappointment. Instead, the Turnaround Lead and all team members should show gratitude for the attempt to improve the Program. The Lead must work with the team to garner any lessons learned from the improvement attempt, then forget this idea did not work and move on.

The team must be encouraged to continue to innovate and improve the program's success with reaching the Turnaround Commitment in any way they can. Team members should be left with the message that it is better to bring forward ways to improve the program and fail than to not try at all.

Finally, a point I believe is very important: there really are no "crazy" ideas from any team member. I never heard an idea during my entire career that did not have *some* merit. If you're leading a Turnaround, you're working with extraordinarily enthusiastic and capable people. Sometimes when a dedicated team member stretches heart and soul to find a way to make the team more successful, they may overlook one or more deficiencies associated with the new idea; but the essence of the idea may be valuable if these deficiencies are addressed. The attempt to create and innovate must be cherished by management, never belittled. It truly is a privilege to work with team members who are not only doing their job well, but are also on the lookout for better ways to achieve the Turnaround Commitment.

13.5 Chapter Highlights

- Leadership should ask subject matter specialist "What If?" to inspire solutions.
  - Encourages forwarding improvement suggestions from all team members.
  - May provide an example of an inexperienced suggestion spawning progress.
- Leadership must share innovative suggestions with the program, even if they don't succeed.
  - Reiterates that valuable innovations sometimes result from many attempts.
  - Reemphasizes value of continuous improvement to achieve the Turnaround Commitment.
- Share specific benefits of successful innovations.
  - Demonstrates value of innovations.
  - Encourages team members to forward new ideas.
- Never punish for suggestions that do not work!.
  - Do not imply disappointment or discontent.
  - There are no "crazy" suggestions.

Chapter Fourteen

Honest Tracking

While executing a program plan, it is imperative to know the status of the program's progress and exactly what work remains to be completed.

Even if the program team members are putting in the time it takes to complete their work when promised, there can still be unforeseen issues that arise. For example, required materials may not show up on time from a supplier; there may be more than predicted incompatibilities among elements when integrating the product; more than anticipated test failures may occur. How does the Program detect these unanticipated setbacks early and incorporate solutions into the Program Plan?

14.1 One Step at a Time

There is an important lesson that good leadership learns early and must remind themselves of frequently. That is, unanticipated issues will occur during a project or program and they must be resolved thoroughly *without losing the conviction* that the committed outcome will be achieved.

I will never forget a discussion during the last hour of a manager training class I attended. The instructor displayed a beautiful picture of a skier standing at the base of a snow-covered mountain. The summit was in view in the background and was thousands of feet above the starting point. This skier had successfully skied to the top of this mountain. He had just one leg.

The skier was asked how he got to the top with just one leg. He simply answered, "One step at a time."

- **When you cannot foresee all the steps to be successful.**
- **When you know there will be unanticipated problems to solve.**
- **When this is the first time anyone is doing this!**

Figure 14.1 Leading any big project or program, especially to recover it from near failure, can seem impossible. Remember the skier with one leg.

When constructing and executing a program plan to achieve a turnaround commitment, the task may seem daunting. It may seem impossible to comprehend all the steps needed to execute the plan in a single thought. The whole team knows there will be problems along the way; they plan to solve them quickly, but they don't yet know what they will be. The size, importance, and uncertainty of the task may seem overwhelming. How will the team survive? The answer again is one step at a time (see Figure 14.1).

For the critical importance and high pace of a turnaround program plan, the steps should be at least one a day. Each day the program must precisely review its development status and the plan for the next day and verify that the turnaround commitment has not changed. The Program must predict and create plans to mitigate the risks that will be present during the next reporting duration, resolve the issues that may have occurred, and evaluate the risks of taking alternative routes if one or more may become necessary.

It is important to remember that the Turnaround Team must not become disheartened or allow the uncertainties and fears that result from daily program setbacks to dampen their enthusiasm or dilute their conviction that they will achieve the Turnaround Commitment. It is an important role of Program leadership to not allow this fear to happen.

14.2 Thank Goodness for Schedulers!

Precise knowledge of the current Program status, with calm objectivity, is necessary when determining the progress of the Program. All tasks must be assessed. The impact of an issue in one task on the other tasks must be evaluated to determine the total impact on the Turnaround.

Task leads, and the Turnaround Lead cannot develop this status themselves. The first priority of these leaders is to succeed with the tasks they are assigned and not give equal attention at this time to the other program tasks. The Turnaround Lead and the rest of program leadership have the additional important role of keeping the Program sold and constantly improving schedule performance. It is difficult for them to also uniformly analyze current schedule performance and predict its implication on future program performance.

Therefore, it is very important to have one or more independent professional schedulers canvas the task leads and subcontract managers to determine the status and problems with each of their detailed task schedules (see Figure 14.2). The scheduler and/or schedule team lead is responsible for collecting daily schedule and risk status and, with the task leads, laying out the planned work for the next reporting duration (usually one day). In addition, they must independently report to the program the likelihood of the various risks becoming issues, their potential impacts if they become issues, and avoidance and recovery

- **Uses consistent tracking process and objective progress metrics.**
- **Has best perspective of program schedule impacts based on knowledge of task interdependencies.**
- **Summarizes pros and cons of schedule status quickly for leadership.**
- **Usually most forthright with "bad news" so issues can be promptly addressed.**

Figure 14.2 One or more good professional schedulers are usually essential to accurately track and direct any program, especially a program turnaround.

work on all risks and issues. The scheduler(s) may work with the Program Risk Management team to gather data for their analysis.

The Program's schedule team can consist of from one person with minimal training working part time for a small turnaround to a team of professional schedulers for a large program. The size of the group will depend on the size and urgency of the work required. Regardless of program size, this function must:

- Be empowered by the Turnaround Lead and enterprise management to evaluate the program schedule status and schedule risks. They must be permitted to do this independent of any influence from the Turnaround Lead and without any censorship or avoidance actions from any element of the program team, customer, subcontractors, or the enterprise.
- Report program schedule status and risk, which is derived from measured data. Report on any lack of requested data.
- As needed, update task-level and program-level detailed schedules for the next reporting duration (often one day) and for the completion of the Turnaround Commitment.
- Identify any impacts on program-level schedules resulting from individual task issues.
- Identify all schedule risks on the critical path.
- Identify schedule threads currently off the critical schedule path that may be on the critical path later.
- Use a consistent process for performing all their schedule work. This process may be derived from the command media or other sources of enterprise work standards for the enterprise.
- Identify any other significant development or schedule risks not included in the above.

The scheduler(s) must present their reports and schedules to the entire leadership team when program status is being reviewed.

14.3 Multiple Books

The approach I am about to discuss is frequently used in good programs with a very high degree of success. In fact, seasoned team members will sometimes suspect it is being used to manage their work yet they meticulously follow the schedule guidance they are provided.

I am referring to the value of Turnaround leadership developing two schedule and cost plans. The work to be completed with both of these plans is the same; however, each plan is created with a different set of assumptions (see Figure 14.3).

- **Plan 1 – Idealistic Cost and Time to Complete:**
  - **Assumes ample staff.**
  - **Assumes facilities on time and complete.**
  - **Assumes supplies delivered complete and on time.**
  - **Assumes no test failures.**
- **Plan 2 – Realistic Cost and Time to Complete:**
  - **Consults actuals from past similar programs.**
  - **Incorporates typical delays addressed by experts.**
  - **Incorporates margins added by program, enterprise, and suppliers.**
  - ***Manage* via Idealistic – Plan 1.**
- **Promise Realistic – Plan 2.**

Figure 14.3 Managing a program or project with multiple books is a good way of achieving milestone commitments with little last-minute panic.

14.3.1 Plan 1: Optimistic Cost and Time to Complete

This schedule and cost plan assumes that no issues or failures will occur during the execution of the plan. It assumes that all needed supplies, information, and other inputs to the Program arrive on time and are complete. It assumes enterprise and contractor facilities are available on time, are complete, and work as needed the first time. It also assumes that no failures are found during testing. Finally, it assumes that all designs and inventions used by the Turnaround Program perform completely as promised.

A program leader understands that this schedule and cost plan is not a realistic forecast of the work needed to achieve the Turnaround Commitment, as it assumes nothing goes wrong. But during times when it is imperative to put the Program's most optimistic outlook forward, this is the plan to show. No one can say it is wrong until an unanticipated problem occurs.

14.3.2 Plan 2: Realistic Cost and Time to Complete

This schedule and cost plan is tempered by past leadership experience and actual schedule performance data for programs similar to the one being saved. These comparisons will be used to adjust the estimated schedule durations, even if there is a likelihood of schedule reduction due to new and improved processes.

The shrewd manager will not assume that a new development process will reduce schedule duration unless this has been proven with a past application.

Subject-matter experts who have been a part of a similar development program may then further review the resulting schedules. These experts will insert any additional schedule delays they have found to be typical in their experience.

Special attention must be paid to any portion of this realistic schedule in which estimated durations cannot be based on past development experience. The schedule and cost margins added to these areas must be higher than for areas supported by performance actuals due to the high uncertainty of the size of the margins needed.

14.3.3 Manage Optimistic, Promise Realistic

The optimistic schedules (Plan 1) are used by leadership to manage the Turnaround with the new Program Plan. The task team leadership and program schedulers must track to this schedule.

As discussed earlier, this is not deceitful. No schedule issue has yet been encountered, and so these schedules are accurate. If a test failure occurs, if there is a late delivery from a vendor or subcontractor, or even if the basic design of the program product is in error, there are processes in the Turnaround Plan to mitigate these sources of schedule delay. It is assumed that the whole program team will make the extra effort needed to keep the Program on this schedule.

This approach must be used to keep a Turnaround Team on a challenging schedule. All team members must appreciate that the success of the Program requires more than providing the required product for the Turnaround Commitment—it also requires solving unforeseen problems along the development path and still deliver the complete results on time.

But the Turnaround Lead must *promise* the customer and enterprise executive management that the Program will achieve Plan 2, the realistic cost and schedule plan. Even if managed well, the optimistic schedule and cost plans may slip after a vigorous effort to preserve it. But it should not slip past the values promised in the realistic cost and schedule plans. With this approach, the Program will often be completed using fewer resources and/or delivering ahead of the schedule dates promised in the realistic plan.

14.3.4 The Tale of Three Books

I must tell of a time in which I kept not two but *three* books. This was for a program that had to build over 100 copies of a highly complex unit and deliver each one on time. We used small, highly productive, multifunctional product teams to accomplish these deliveries.

Toward the end of our delivery schedule, we were offered financial incentives from our customer to deliver our last units earlier than agreed. Thanks to the wisdom of my business team, we established a system of three cost and schedule books to manage the program.

The plan with the shortest schedule durations and tightest cost constraint was the one we used to manage the task teams doing the work—call it plan A. A second plan (plan B) with slightly longer schedule durations and relaxed cost constraints was what the functional managers and task leads worked to. They appreciated that Plan B had margin in the schedule and cost constraints they were managing to with their teams in Plan A. But a third plan (plan C) with still more schedule and cost margin was what the Program Manager promised to the customer. The schedule milestones in this third plan would achieve 100% of the early incentive fees offered by the customer.

Of the over ten deliveries under this three-tier plan, every delivery earned 100% of the promised early delivery incentive. The customer was very pleased the program team's successfully achieving all their early delivery commitments.

Yet the program team seldom worked long hours or endured unusual pressure to achieve the milestones in plan A. They knew from the start that this was the box they had to fit in. Some of the team members may have surmised that there were margins around their plan, but as good professionals, they committed to the numbers and dates they were assigned, and they achieved them.

On occasion, a completion date in plan A slipped despite the best efforts of the team to solve an issue. The Program and enterprise treated the slip with the same seriousness as if it was a slip to the promised date to the customer. But only the program manager, the program's business manager, and the customer knew what actual delivery dates and maximum allowable costs were promised.

14.4 Keep Watching the Flank

As discussed earlier, during the heat of keeping a Turnaround Program on schedule, the emphasis will often be on the task thread in the plan that takes the longest schedule time to complete, often referred to as the critical path. Most of the program effort is usually devoted to adhering to the milestone commitments for the tasks on this critical path.

The problem is that while the Program is focusing on the critical path, the issues developing in other tasks that are close to the critical path may not get sufficient attention. As a result, as the tasks on the critical path are completed, other tasks may be found to be delinquent as a new critical path emerges. This often leads to a program jumping from one critical path thread to another, seemly always in a defensive state of solving unexpected issues, instead of systematically identifying risks and planning solutions before they become issues.

The answer to this problem is, instead of working just those tasks on the *current* critical path, resources should be devoted to the tasks that will move up to the *next* critical path. No task lead should allow the progress of their task to fall behind in their schedule, believing their output will be needed later than planned due to delays in other task teams.

I have taken over programs in which different development groups, geographically separated, had unwittingly developed what I called "schedule slip ping pong." One group would back off the pressure to achieve their next milestones seeing that another group on the critical path was falling behind. But when this schedule slip was fixed, the first group then found themselves back on the critical path and the group that had been on the critical path relaxed their schedule drive. The net effect was that these back-and-forth slips were endangering the completion date of the whole program.

The solution was to hold each group to the schedule they had, without regard to the schedule performance of the other group. The teams were reminded that just because some schedule was *starting* to slip did not mean it was *going* to slip. The job of each program team was to insert corrections and do extra work to meet or better their committed milestone dates. After this policy was instituted, the achieved completion dates of these two groups actually moved ahead of their promised completion dates.

There may be a situation where the program does not have enough resources to provide ample scrutiny of all program tasks to check for high-risk ones. This may occur, for example, while a program is late in achieving its staffing needs. During this time, the tasks on at least the task thread that would be the new critical path if the current critical path were eliminated should be given high attention so they are completed on time. Most computer based schedule tools will easily show the next critical program path if the current critical path is shortened.

An important warning of an upcoming task that may be riskier than anticipated may come from someone not on the Program's leadership team, such as program team members, customer members, subcontractor members, and other parts of the enterprise. If someone approaches a Program lead and wants to share their concern about an issue on the Program or a task they believe may fail, they must be listened to carefully! This can be the best "heads up" the Turnaround Program leadership will receive about a subtle issue that may set the program seriously behind.

14.5 The Common Fallacy of "Reuse"

The promises of reuse can be very compelling. "Why design something from scratch if it has already been built?" "Using what exists makes sense." "Someone else paid for its development and testing." "It has been deployed and working in the real world for a while. It must be super reliable!"

So let's review the different places in the typical development cycle in which reuse might be applied.

14.5.1 Reuse the Design Concept or Approach

This reuse can greatly shorten development time when developing something new. The Soviet Union's development of the atomic bomb was shortened by almost a decade by simple pencil sketches on two pieces of paper showing the way the United States had chosen to initiate a chain reaction. The Wright brothers applied the principles of efficient propeller design developed years before they build their propellers for the Wright Flyer. When there are many design alternatives that might work, learning that one has been used and proven to work can save huge amounts of product development time and cost. There are often legitimate savings from this kind of reuse.

14.5.2 Reuse Successful Detailed Design or Finished Product

Here is where often a reuse mistake is made.

A successful design is one that has been used to create a finished product. This product must have been tested for it to have been proven to be successful. For a program that requires an extensive documentation trail of all the development steps, there has been much money spent to document the start-to-finish legacy of this successful product.

However, and this is a big "however," once one line of code is changed in a software program, one fastener is changed in a mechanical structure, one sample group is changed in a survey, etc., the wrapper is broken for the completed product. Even with a minor change, the entire development process must be reviewed and, usually, performed again to verify that some other part of the product has not been affected by the change. In most such cases, there is little or no value gained from reuse.

This sounds counterintuitive. If you change only 10% of the lines of code in a software program, you can't reuse the other 90%? Can't you save 90% of the cost of starting from scratch? If you increase the thickness of a wing spare in the new airplane to allow a heavier payload, can't you reuse most of the rest of design of the airplane?

The answer is that most of the time spent building a successful product is in establishing the design requirements and then testing the resulting product.

For example, for high-quality embedded software, only 2% to 3% of the cost is spent actually coding the software. The rest of the cost is in developing the software requirements, designing the software, and testing and integrating the resulting code.

When you break the wrapper for any detailed product design or finished product, you must review and usually re-perform all the requirements flow down and evaluate the need to change any derived requirements. You must update the detailed design of the product and check for the effects of the change on the rest of the original product.

Also, most, if not all, of the product testing must be repeated. This is because, when even a small portion of a design is changed, it is almost impossible to be certain what other portions of the product or system have been affected. Complete retesting must be performed to make sure any potential adverse effects from the change have been found. As a result, all requirements-related, design-related, and test-related documentation must be updated (see Figure 14.4).

Unfortunately, this fallacy of reusing portions of an existing detailed design or working product without further work to update the documentation for the whole system and retest pertains to all classifications of products, including engineering, studies, analysis, business models, production sequences, management procedures, etc.

- **Most value lost when existing wrapper broken:**
  - **Even a little!**
  - **Must revisit requirements analysis, design, test planning, testing, and documentation.**
  - **Applies to all hardware and software products.**
- **Hope of hardware and software "chips":**
  - **Chip function must be thoroughly documented.**
  - **Form, fit, and function must be meticulously frozen.**
  - **User confidence must be high.**
  - **Successfully applied for digital electronics.**

Figure 14.4 The loss of almost the entire investment of a hardware or software product development work after its wrapper is broken may be counterintuitive, but even a slight change to one element of a product may affect its overall performance. Therefore, reuse of developed product seldom saves significant cost and time. This also pertains to studies, analysis, and planning products.

14.6 Building Component "Chips"

There are ongoing attempts to build elementary functions into fixed components that are then integrated to build up a system. This has already been done for digital electronic components—we sometimes call these "chips." Each electronic chip has a product data sheet that describes in detail its physical size, its power requirements, the fixed function that it performs, and all the electrical and time characteristics required to enter information into the chip and interpret the result computed by it. These chips often abide by a common standard that describes the electrical and signal timing requirements that must be followed to allow the different chips to communicate with each other.

Software and other technologies are trying to follow a similar approach. For example, one of the goals of object-oriented software design and the application of polymorphism is to have applications' unique software logic automatically call up the basic software's supporting functions from a library of "software chips." This could reduce the development risk and total cost needed to build and test new software logic.

The development and use of standard chips is being evaluated for other products as well. The benefits of using the chip concept is still emerging. But for any application it is only successful if there is very high confidence in the completeness and accuracy of the documented performance of a chip that is made available.

14.7 Chapter Highlights

- When it seems impossible, proceed "one step at a time."
- Foundation.
    - Carefully establish the Program Plan.
    - Track it daily with appropriate metrics.
- Independent scheduler(s) are highly valuable.
    - Use consistent process and measurements to track progress.
    - Thoroughly assess program impacts based on thorough knowledge of task interdependencies.
    - Are forthright with bad news.
- Program management with two books.
    - Plan 1: Best-case completion.
    - Plan 2: Realistic completion.
    - Manage with best case, promise realistic.
    - A story of three books.
- Watch the flank.

  - Make sure tasks on the next critical path are on time in case the current critical path is shortened.
  - Record and evaluate risk warnings from all familiar with Program Plan.
- Common fallacies of reuse.
  - Savings are often overestimated.
  - Breaking the existing "product wrapper" invalidates almost all requirements analysis and test work.
  - There is hope of building reusable component "chips."

Chapter Fifteen

I Thought I Understood Software!

The growth in the application of computers in the new products we create is amazing. These computers often reside and operate in the product they are controlling and are referred to as *embedded computers*.

These computers have increasing roles in our lives, from controlling home appliances, providing more efficient use of fossil fuels, and providing safety in our cars, to actually driving our cars and flying airplanes and spacecraft. With software, these computers are providing greater portions of the total functionality of today's products; plus, software-based utilities are now a valuable tool when planning and leading turnaround projects and programs.

But the task of understanding and developing software is still in its infancy. Despite the proliferation of personal computers and social media, most of us have never written a software program. Universities have only recently provided degrees in computer science. Electrical engineers are increasing processing speeds and memory capacities in computer hardware at unbelievable rates, but what is the status of the engineering of software for these new high performance machines?

Moreover, most program leadership has little knowledge of or background in software technologies, yet are highly dependent on it. What more should a program leader know about software development?

15.1 New Guy on the Block

As mentioned, the role of software in our products is constantly increasing, and this trend is not going away. To some of us, it may seem like software and computers have abruptly invaded our lives.

It should be pointed out that software lives mostly by a set of man-made rules and not laws established by mother nature, which is traditionally the source of constraints for the classical sciences and engineering. When a computer program fails to perform the intended algorithm, it is usually because the software developer broke the user rules established by another human being and not the constraints of our natural world. These rules usually must be followed perfectly. There is no approximate or so-called "first order" or "back of the envelope" manifestation allowed for software applications. This trait is often surprising and even exasperating for people with backgrounds in the classical engineering and science disciplines when they try to develop software for the first time.

Because computer hardware design is evolving with increased performances so quickly, these strict and seemingly arbitrary computer application rules are constantly changing, which makes understanding software even more of a chore. The highly sought after Job Control Language and PL-1 programmers of the past are now totally obsolete. Structured software design has given way to object-oriented software design. Software tools that apply current development methodologies and automatically generate software code are rapidly being developed. Remaining a competitive software engineer needs the additional dedication of staying up to date with the rapidly changing software technologies and tools that are available.

What sometimes adds to the mystery of software engineering is the proliferation of informal colloquialisms among developers. Terms such as "pseudo operations" and "orphan discretes" have clear meanings for the impassioned software engineers communicating with each other, but they often leave the non-software team member perplexed and sometimes doubtful that these software people are serious about their work. Let me assure you, they are!

15.2 Software Is Not All Created Equal

A common and unfortunate mistake made by many program leaders without a software background is that they assume all software is the same. This is a serious error. It's unfortunate that task leaders and managers sometimes hire software professionals just by their affiliation with software, without regard to what applications specialty their experience is with—sometimes mistakenly hiring as many software engineers as needed to fill a headcount target, regardless of what their domain specialties are.

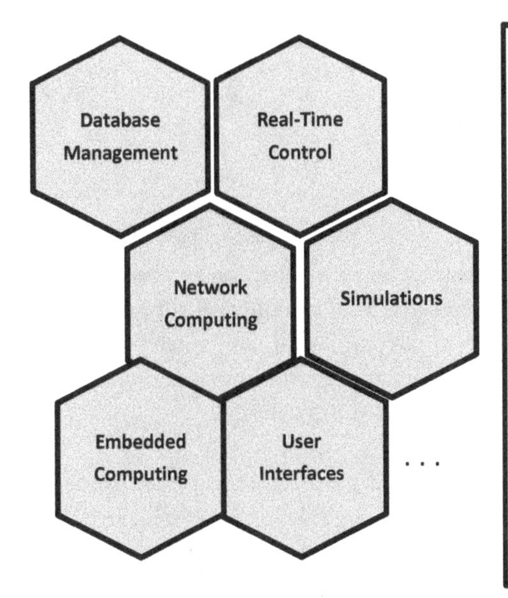

Elements Unique to Each Domain Include:

- **Operational Concept**
- **Typical computer performances**
- **Data interface architectures**
- **SW Operating Systems**
- **Programming Languages**
- **Electronic technologies and functions used**
- **Interfaces that need to be controlled**
- **Commercially available SW**
- **Commercially available SW development environments**

Figure 15.1 It is a mistake to think that all software engineers are the same. Instead, they usually have a specialty in one or more application domains. Here are examples of some.

The fact is that modern software engineers usually develop specialties in very specific application domains (see Figure 15.1). These specialties include embedded real-time software, database management software, terminal user interface software, network software, simulation software, and many more.

Each of these applications has its uniquenesses, including the need to understand the engineering disciplines they are supporting, the typical computer hardware architecture the software will be operated in, the typical algorithm designs best suited to be implemented in this domain, the computer operating systems available to execute the applications software, the best programming language(s) to use, the typical operational concept for the software they are developing, and much more.

A new software engineer usually selects one domain specialty and develops a specific understanding of factors such as those mentioned above. In addition, they eventually develop an experience-based knowledge of the typical development errors and bottlenecks to watch out for in their application domain and how to quickly work around them. They understand the content and purpose of the typical development milestones needed to plan and develop the software for their domain.

When assembling or revising a team to build software for a program turnaround, it is especially important that leadership recruit talent that specializes in the application domains being developed by the program. Failing to do so is often one of the reasons—sometimes the major reason—a development program fails.

15.3 How Can I Touch Software?

You can look at an automobile and estimate what its purpose and performance is before it moves an inch. Similarly, you can hold a report on the success of a medical treatment, or look at an electronic circuit board, and at least say, "Here is a product. I can see what its approximate scope is and how it is divided up. I can make a first-order assessment of what it does. It's tangible."

So how do you do this with software?

Some will say you can hold the programming listing. If you understand the programming source language, you can get some idea of what the purpose of this software is. But you will quickly be lost in the details of the syntax of the language. You are only holding a description of what you want the eventual software to do. It is not the actual software.

If this source code is written in a structured language like C++, you will see some division of the software logic. However, it may not be certain if the structure you observe is for the sake of partitioning the logic being implemented or because it is required by rules of the programming language syntax.

Good programmers will insert descriptive comments in their source code to guide the reader through the intended logic and to make it possible for the reader to better understand the purpose of the code and the division of the computing tasks. These comments are like margin notes on the outline for a report or design notes on a schematic for an electronic circuit. But they are not the actual report or circuit board.

The product called "software" is really a high-density accumulation of electrical states in a computer's memory. Each state may have one of two electrical values. Each one of these states may be individually interrogated or changed to the other state. Each state is, of course, called a *bit*.

We can hold a computer memory device in our hands but cannot see the electrical states of the bits. And if we could, these states would have no meaning until a computer interrogated them and decoded them in some specific sequence. Unlike other things we make, software has no tangible meaning until it is executing.

The important point I wish to make is that software is more removed from having any tangible qualities than other sciences or technologies. Any tool that can be used to make software more tangible to humans is of very high value. As a result, mapping the required software functions to appropriately named software programs, for example, must be easy to understand by everyone on the program, not just the software team.

A clear understanding of what functions each piece of software performs and how these functions interact is imperative for the software engineers, for Program leadership, and all members of the Program Turnaround. Otherwise, the Program team will lose track of what functions they have provided in the software and in what software programs these functions reside. Failure to precisely maintain this knowledge often becomes a major contributor to the failure of a development program.

15.4 Software Cost Myth

Software cost is often measured and evaluated as the total labor hours spent per line of code completed. There are a host of very good computer-based software cost models available, such as PriceX and COCOMO, that estimate the cost for developing software for a specific application. These programs ask the operator a series of questions that describe the software, the size of the task, the size of the software development team, the software design complexity, and much more. The cost estimates generated by these tools are distributed among the standard software development phases, including requirements analysis, preliminary design, detailed design, coding, unit test, system test, integrated test, and maintenance. These models are largely based on actual software development costs accrued by recent programs that have been successful. These models are commonly used by software development teams to crosscheck their cost estimates during planning.

For those of us who do not write software, we may mistakenly think that most of the software development work is done sitting at a computer typing in program code. But it turns out that writing code accounts for less than 8% of the total cost of developing software for non–real-time applications that may execute on desktop and office-type computers. For the high-speed, high-reliability software that runs in embedded real-time computers such as automobile transmissions and aircraft autopilots, coding this software accounts for only 2% to 4% of the total cost! The rest of the development cost is approximately 35% for software requirements analysis and design and approximately 55% for software testing (see Figure 15.2).

It is therefore important to note that if software development cost must be lowered, then lower-cost methods to perform software requirements flow-down, design, and testing must be found. These items are the major sources of cost. Figure 15.3 lists the factors that drive most software cost estimating programs, such as COCOMO.

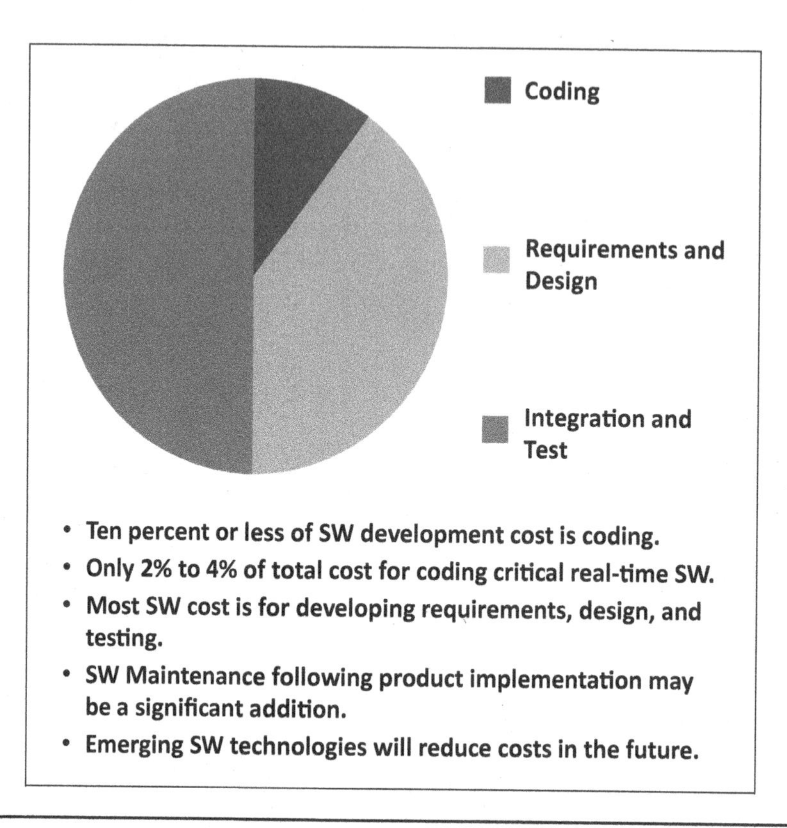

Figure 15.2 This is the typical cost allocation for the development of software for all domains. Total cost will vary for different applications, but coding is a small part of it. Automated software development environments and advanced coding languages are reducing total cost.

15.5 Software Programs That Make Sense

The software used in a most projects and programs is usually broken into some number of software programs. For large amounts of software, this first layer of decomposition may be referred to as *software subsystems.* These subsystems are then usually broken down into software programs. Smaller total amounts of software in a product may be simply broken directly down to programs.

Any non-software person who is shown the decomposition of a program's software task into subsystems and/or programs should be able to point to the one program that a major software function executes in. They should be able to accurately infer this from the name of each program, as each program's name should clearly describe all functions that are executed in it.

- **Personal Factors**
  - **Application experience**
  - **Programing language experience**
  - **Experience using virtual machine**
  - **Capability**
  - **General SW experience**
  - **Continuity developing SW**
  - **Platform experience**
  - **Experience with language(s) and tools**

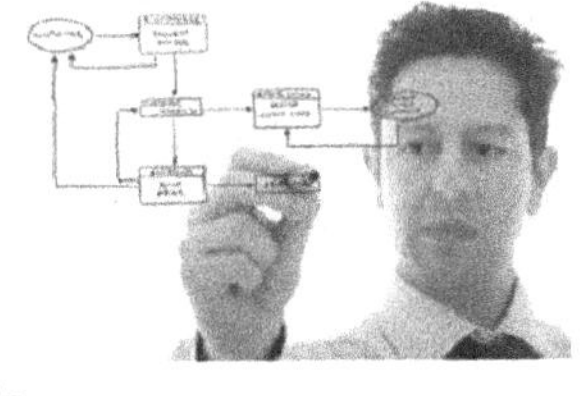

- **Product Factors**
  - **Required software reliability**
  - **Database size**
  - **Software product complexity**
  - **Required reusability**
  - **Match of required documentation to lifecycle needs**
  - **Product reliability and complexity**
- **Platform Factors**
  - **Execution time constraints**
  - **Storage constraints**
  - **Volatility of the operation of the virtual target, if available**
  - **Volatility of the operation of the target**
  - **Difficulty with using the target**
- **Project Factors**
  - **Use of software tools**
  - **Use of modern programing practices**
  - **Required development schedules**
  - **Software/data protection**
  - **Multi-site development**
  - **Requirements volatility**

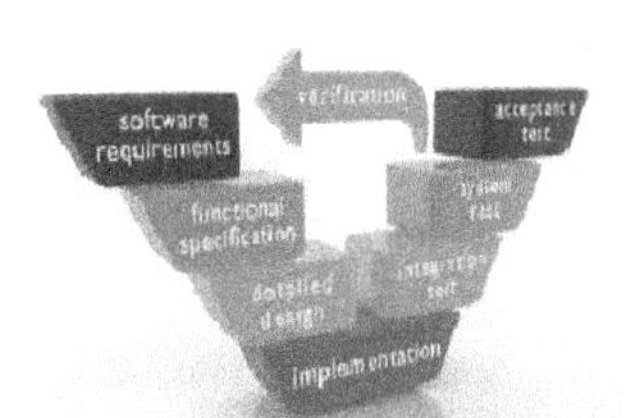

Figure 15.3 These are the major factors used in the most popular tools for estimating software cost. They are the typical cost drivers.

Unfortunately, for many development programs, the lack of well-engineered software divisions and the choice of vague program names do not allow this. The software programs may have been divided up without the use of a proven software design method.

New software functions may have been added to existing software programs that have nothing to do with the purpose or functions of the existing program. These programs may become what I refer to as "garbage can programs." As a result, new software developers on the program team cannot tell where many of the basic functions reside in the software. Worse than that, the development of poorly designed software may reach a point at which even many of the long-time software developers on the program cannot tell where the basic software functions reside! I have assisted big and important system development programs with exactly this problem.

When the software being developed becomes this disorganized, there are sometimes just one or a few developers who remember what functions are performed in each poorly designed program. These individuals often acquire a dubious title, such as "software guru." At this point the development program is in trouble.

When I was a junior engineer, I knew an excellent project manager who had little experience with software tell me, "Software is like a pie. You can cut it up any way you want, and it is still the same software." This is wrong—you can see and hold a pie.

I am sure this same person went home and walked through a door to enter their house, not a window. But why not make the window a little bigger and call it a door? You can improve efficiency by having two uses for that hole in the wall. Isn't it still the same house?

15.6 Need for Independent Pieces

The answer is that we as humans have the need to divide big things into component pieces such that each has a specific, easy-to-describe task and that each task has minimal interdependencies with the neighboring tasks. The door is for entering and exiting the house. The window is for letting outside air in. Clear single roles, little interdependency.

As discussed earlier, independent software programs are achieved by having each program contain highly related and closely interdependent functions (maximum functional cohesion) while requiring little communication among programs (minimum coupling). The techniques of structured analysis and structured design use a hierarchy of definitions for each of these two qualities to create a software design that achieves this highly independent partitioning (see Figures 15.4 through 15.6).

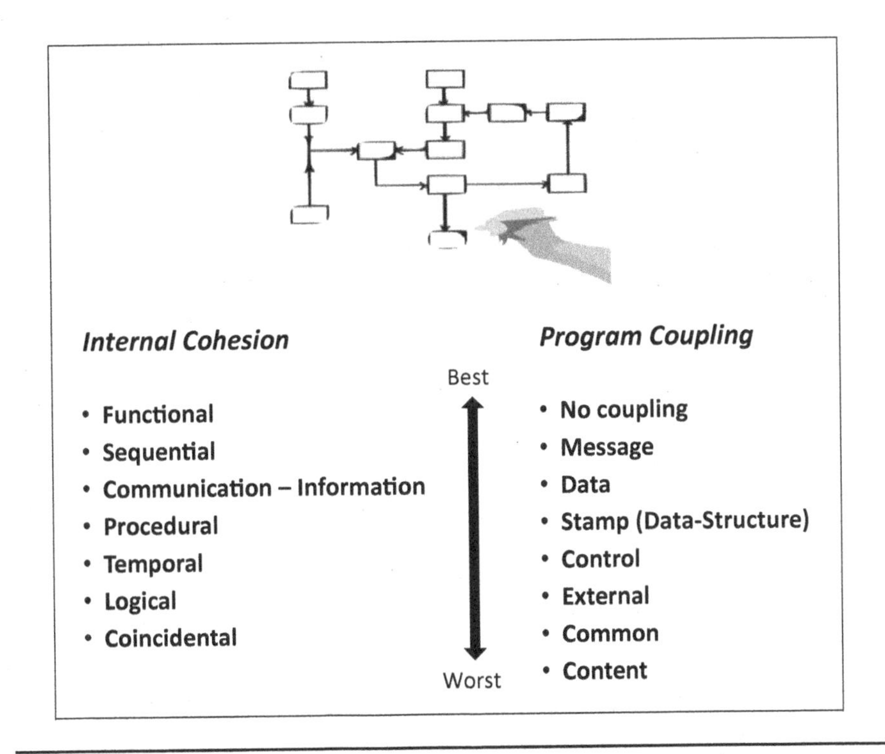

Figure 15.4 Creating a software design with programs and subprograms that are highly "independent" is essential to keep software development and maintenance costs low. Maximizing functional cohesion, minimizing data coupling, and choosing descriptive program names helps establish this.

Please note that these traits of maximum functional cohesion and minimal information coupling are useful guides when breaking down a turnaround program to its most independent organizational pieces or deciding how to distribute program work to geographically separated work sites. These applications require less process rigor than software design does, but the principles of high cohesion and low coupling provide a good crosscheck of the divisions selected.

Object-oriented design goes further by encapsulating a description of the program function and all external data interfaces in one place in the program, using a common formant. This ability is facilitated with the object-oriented language syntax, such as that used in Ada, C, and C++.

A further goal of object-oriented design is to simply write the applications software at a high level of abstraction and rely on an existing library of standard software programs to perform the more basic computing functions. For example, why write new software to transform a locus of points in a Cartesian coordinate system to a polar coordinate system if this logic has been programmed and used successfully many times before? The linking of required software

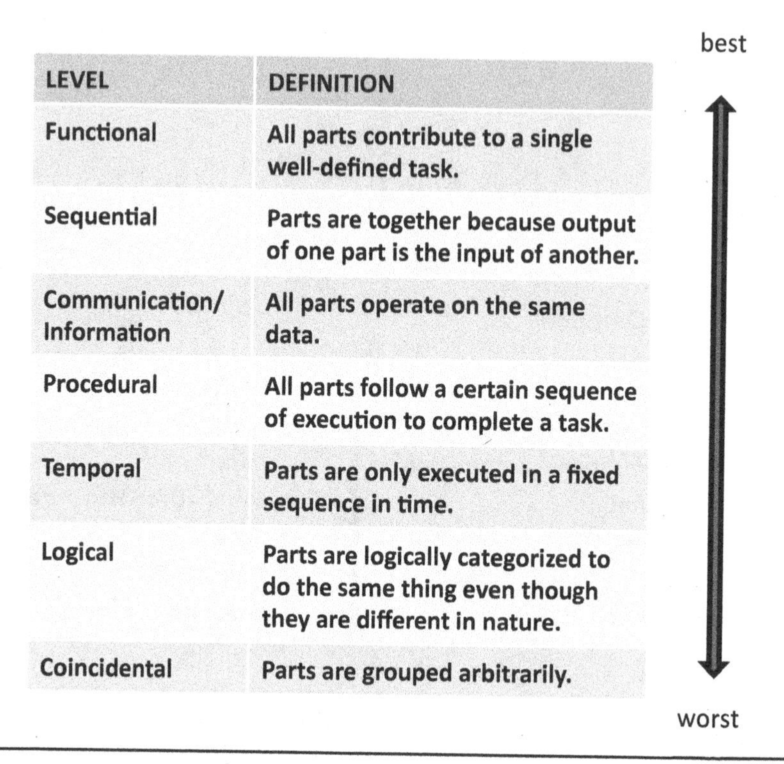

LEVEL	DEFINITION
Functional	All parts contribute to a single well-defined task.
Sequential	Parts are together because output of one part is the input of another.
Communication/ Information	All parts operate on the same data.
Procedural	All parts follow a certain sequence of execution to complete a task.
Temporal	Parts are only executed in a fixed sequence in time.
Logical	Parts are logically categorized to do the same thing even though they are different in nature.
Coincidental	Parts are grouped arbitrarily.

Figure 15.5 During design, each software partition is adjusted to obtain the maximum level of functional cohesion. This analysis is also useful when deriving the best work breakdown for the Turnaround Program and when assigning program work to different geographical sites.

functions in the object-oriented software development system is accomplished with a software-driven search in large object-constructed software libraries for the software functions needed. This software doing the linking verifies that the routines chosen provide the performances the software engineer specifies, and then includes these routines in the final build of the software. This object-based linking function greatly reduces software development time, test time, and test risk.

As mentioned earlier, an important ingredient to ensure that all team members understand what the functions are that a software program provides is to give each program a name that describes specifically what it does. Sometimes the name can be a short phrase. Names for programs such as “Partial Database Update” and “Altitude Update” are vague. Names such as “Compute Start and End Dates” and “Compute Barometric Altitude” are more explicit and result in fewer errors occurring in a development program or project.

best

LEVEL	DEFINITION
No Coupling	Modules do not communicate with each other.
Message	Modules exchange discrete state information in messages.
Data	Modules exchange data via direct communication.
Stamp	Modules use data in a common database.
Control	One module can control the functions in another.
External	Modules share an externally imposed data format, communication protocol, or device interface.
Common	Modules share the same global data.
Content	One module can change the internal workings of another module.

worst

Figure 15.6 During design, each software partition is adjusted to obtain a minimum level of data coupling between them. As for the evaluation of cohesion, this analysis is useful when deriving the best work breakdown for the Turnaround Program and when assigning program work to different geographical sites.

15.7 Find the "Bugs" Early

The cost of repairing defects found while testing software increases exponentially as test failures are encountered later in the test plan. This is because:

- More enterprise facilities, test hardware, and test software are idle waiting for a software defect to be repaired as testing becomes more integrated.
- More staff is needed to maintain test facilities, set up and run the regression test, and analyze the failed results when the software is more integrated.
- There are more software and hardware components being tested at higher levels of integrated software testing. Therefore, finding the cause of the test failure often takes longer.

- There are fewer test facilities to support the more integrated levels of software testing. Therefore, there is less opportunity for other testing to continue in parallel when a failed test is being analyzed. Software test failures at higher levels of product integration result in greater risk of not achieving Program milestones.

For these reasons, it is imperative to design the software test program to find defects as soon in the testing sequence as possible.

The first test of a new software program is often referred to as *unit test* (see Figure 15.7). This is typically performed at the software developer's workstation by just the developer. The test is usually set up and conducted by a common software test program and process used by the Program. The anticipated environment the software is designed to eventually operate in, including target computer hardware, device interfaces, operational scenarios, etc., are usually simulated with software.

All software developers on a program should use the same unit software test tools and follow a standard unit test procedure, usually described in the program's Software Development Plan. The description of the test case, the

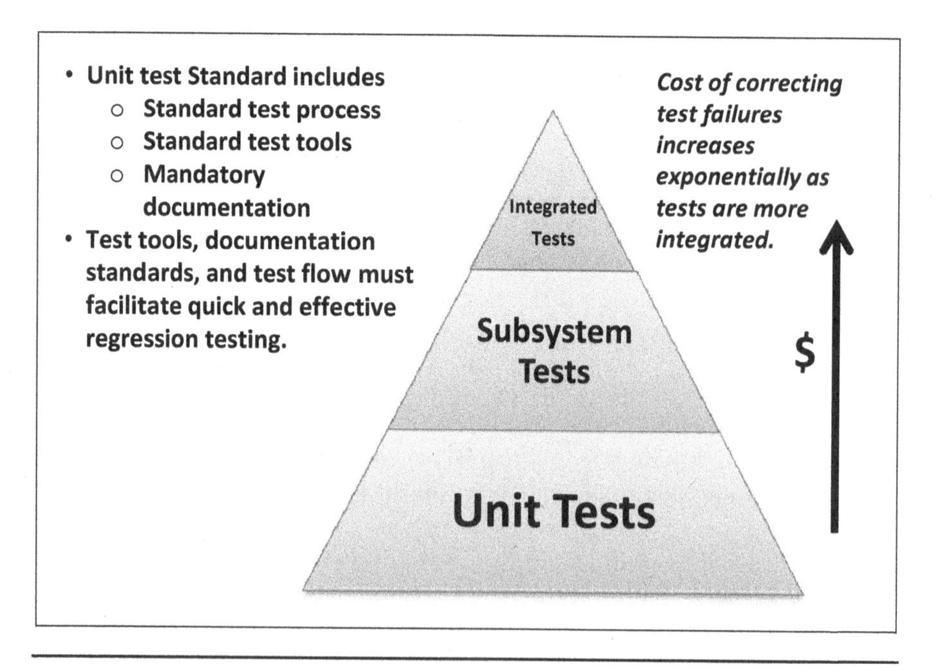

Figure 15.7 Make sure new software programs are thoroughly tested at the unit level before integrating with other software and the system. A standard unit test process must be used and the test set up, process, and results documented. The program must be able to quickly regress to past unit tests when an anomaly occurs during integration.

test setup, and the test results should be stored in the program configuration management system. This approach helps ensure that thorough software testing is done at the unit level and that a unit test can be accurately rerun if the software unit is suspect during the search for root cause of a product failure. These regression tests are often performed when there is a test anomaly encountered at a higher level of product integration.

I have led program turnarounds in which software unit testing had been found to be inadequate. One example was for an embedded software application for which unit testing was very ad hoc, with no program standard. The first formal testing this software saw was in an expensive, highly integrated software test environment. Not only did this environment use a large number of multiple computers and other resources, but each test required many hours to run. Software failure symptoms in this large environment were ambiguous, and the fault causes were expensive to determine and repair.

This program was approaching a series of critical software delivery milestones that, based on past test schedule performance, were forecasted to greatly slip. I led the team to quickly introduce a standard system of unit testing before taking this software to integrated system testing. The results of each unit test were quickly reviewed by a group of peers to make sure the test was complete and passed. As a result, total software test schedule durations and costs were reduced, and all future software was delivered on or before the promised time. In addition, the program now had a controlled set of unit tests to fall back on in the event of a system-level test failure. Not a brilliant process remedy. We were just following the basic discipline for developing good software.

15.8 The Software Ambassador

There are a few frequently occurring observations that should be pointed out before starting this discussion.

First, most Turnaround Leads and program leaders do not have a background in software development. They may be very capable of using popular desktop software applications such as Excel, PowerPoint, Word, etc.; however, they probably never actually wrote and tested applications software, wrote and tested an operating system, or predicted software memory and computational throughput requirements.

Second, many software managers have been promoted to this role for reasons that are justifiable but often not the best. Often a senior software engineer is promoted to a leadership role because of their long experience with software development and possibly their long tenure with the enterprise. They may have a gift for analyzing complex software problems and consistently applying successful solutions.

- **Plan introduction and overview**
  - **Purpose and scope**
  - **Assumptions and constraints**
- **Relationship to other program plans**
- **Referenced documents**
- **Identification of all software and software products to which this development plan applies**
- **Definition of terms and acronyms**
- **System and software architecture**
- **Overview of required work**
  - **Requirements and constraints**
  - **Software products and deliverables**
  - **Project resources and schedules**
  - **Software risks**
- **Project organization and resources**
- **Software development activities**
  - **Software development processes**
  - **Software development methods**
  - **Software development standards**
  - **Reusable software products**
  - **Software types/categories, including operational software, test software, support equipment software**
  - **Software safety, security, and information assurance**
  - **Software engineering environment**
  - **Target computer resources estimates, reserve capacity, and growth management**
- **Software development processes**
  - **Detailed development methodology**
  - **Software development tools and environment**
  - **Prototyping and simulations**
  - **Software requirements analysis**
  - **Software preliminary and detailed design**
  - **Software coding**
  - **Software unit integration and testing**
  - **Software component integration and testing**
  - **Product integration and test support**
- **Software risk management**
- **Supporting information**
  - **Approach to requirements traceability**
  - **Rationale for development methodology and tools**

Figure 15.8 A Software Development Plan is necessary for a turnaround to ensure that developers are using the same processes and tools efficiently. It is the software developer's handbook and can be amended with a process described in the Program Plan. Good ones are often short and concise. Here is a general outline.

But, what a Turnaround Lead needs is a software manager who is very sensitive to emerging software development issues at the program level and knows how to remedy them. This manager should have the ability to fluently communicate the software development steps and associated issues in non-software language to the leadership team. The manager should understand that even if the software development team says they are 98% done with the software, the last 2% of the work usually takes much more than 2% of the total software development time. They should be able to anticipate and respond to the typical software problems that appear during program or project development. These good software managers may have a reputation of almost appearing to "overkill" avoiding typical software risks with effective solutions to keep the new program plan on schedule. They make sure that a Software Development Plan is completed, that it is easy for the whole program team to understand, and that it is rigorously followed. See Figure 15.8 for the typical outline of a Software Development Plan.

The good software manager translates the description of any software development steps and software problems provided by the software experts into terms that anyone on the program can understand. They are the "software ambassador" for the Turnaround (see Figure 15.9).

The Software Manager must report directly to the Turnaround Lead and be on the program leadership team. I discussed in an earlier chapter a recommended

- **Being a senior and experienced SW engineer is not all it takes to be a good SW Manager.**
- **Must have experience with program-specific SW domain(s).**
- **Must fluently communicate SW status, plans, and issues to non-SW team members.**
- **Must have an experienced-based intuition that predicts common development issues.**
- **Makes sure the SW Development Plan is complete and is followed.**

Figure 15.9 The Software Manager must be able to anticipate issues and communicate all software development information in layman's language. They must know what it realistically takes to complete each software development step. Often reports directly to Turnaround Lead.

approach to recruiting the most effective professionals for the program team. This approach should be applied when selecting a software manager for the Program, if needed.

In addition to having the abilities mentioned above, make sure this candidate comprehends the big software picture for the Program and is willing to voice opinions about how it might be improved. Make sure they feel confident in anticipating the traditional software development pitfalls in the domains in which the software for the program is being developed. This understanding should be based on their actual work experience. Make sure they will initiate early recovery action if a software development issue is becoming apparent. Make sure they have both the aptitude for and interest in finding exceptional software talent for the program and will use these professionals effectively.

15.9 "Overkill" Up Front

Software does not offer as much tangible evidence of development progress as the physical evidence created by other products. For example, when a software development group claims they have finished a preliminary design, what work products should be examined to verify this? Is this milestone defined in the Program's Software Development Plan? Has an independent person or group verified that this milestone was actually achieved?

Software milestones must have completion criteria that are thoroughly defined in the Software Development Plan and strictly applied. Completion of these milestones is usually substantiated with specific completed work products. Completion criteria must be rigorously enforced from the start of software development, not left for later. A claim of completion of a software development milestone should be independently verified by a software professional or peer team. The Software Manager for the program must be able to certify with confidence that each software milestone has been accomplished.

The importance of staying ahead of the software development schedules cannot be overemphasized. The Software Development Plan must be strictly followed starting the first day of development. During the life of the Program, issues will likely develop in other non-software program components that often can only be solved with a change of software functions. Yet, often when software comes to the rescue, it is held accountable for any resulting delays in milestone completion. When this happens, it is often impossible for the software team to say they are back on schedule.

This is why it is so important for software development to stay ahead of schedule from Day 1. Schedule margin will be needed for software to solve problems in other parts of the Program later on.

I usually manage program software to be complete with at least a 10% to 20% schedule margin—that is, for the software task to be completed 10% to 20% ahead of schedule. The size of this margin will depend on the complexity of the software, the type of software application, the software size, the maturity of the data system the software will execute in, the record of accomplishment of the software development methodology and management tools used, etc.

15.10 Future Software Jewels

There are a number of developing technologies that may simplify software development and reduce development costs. The following are a few that may be valuable.

15.10.1 Highly Automated Software Development/Test Environments

There is a large amount of data that is generated during software requirements flow down and software design that should be carried through software coding and testing. This data is used to verify that the software created provides the functions and performances stipulated in the requirements. This data includes the type of each requirement, required performance values, names of the software program(s) implementing the requirement, names of interface parameters between the software programs, detailed definition of each interface parameter, operational control states, method of verifying each requirement, etc.

Modern software development tools account for these definitions, and more, throughout the development and testing of the software for the program. These tools are usually executed on one or a network of computer workstations. In addition, these development tools facilitate and usually mandate the use of a specific software development methodology.

These tools check for the complete implementation of the functions and parameters established during the software requirements analysis. They account for parameters and functions defined in the software requirements documentation but not implemented, as well as parameters and functions that are coded with software but do not have a documented requirement. These tools help assure all requirements are tested.

While performing their work, these tools maintain a central database of all functions and data used in the software under development, and they usually provide scrupulous configuration management of this database. These tools often automatically create software requirements documents and software design, code, and test work products in standard formats appropriate to the development methodology being used. These computer-based tools provide tracking of software requirements flow down, software configuration control, management of the software design method, and automatic report generation, saving gigantic amounts of software development time and documentation cost—this while reducing errors during the software development process.

Please note that it is important that the Program's software development team completely understand the software development methodology applied so they can fully capitalize on its intent. Just using an automated software development tool designed to use a certain development methodology does not guarantee that the benefits of this methodology will be achieved.

15.10.2 Code Generators

Even with advanced software coding languages, human effort is needed to convert a logic and/or mathematical algorithm into a software program. The programming language syntax rules must be scrupulously obeyed, or a simple but time-consuming error while establishing the intended computation can be made.

In the past, meticulous hand-coding was necessary to minimize use by the resulting software of computer hardware resources such as memory size, the number of computations per second, and interface data bandwidth. But with the advent of very high-performance microprocessors, this need for high-efficiency software is not as critical.

As a result, software tools are emerging that take a highly symbolic representation of required software logic and convert it directly into loadable object code, completely bypassing the need for traditional coding by a programmer.

As mentioned, the code generated by these products may not use the target computer hardware resources as efficiently as the code created by a good software engineer, but with the high performances of modern computer hardware, this inefficiency may not be important. The development time, schedule risk, and cost saved by eliminating the need for a software coder may be much more valuable.

15.10.3 Off-the-Shelf Real-Time Operating Systems for Embedded Computers

In the past, the development of an operating system (the systems software that, among other functions, tells the applications programs when to run and which computational logic to execute) was done uniquely for each embedded real-time application. This was because making the operating system applications-specific could greatly reduce the amounts of valuable computer hardware resources used by this software. The functions of these operating systems typically include at least the applications software executive (time- or task-driven paradigms telling the applications software when to execute), system mode control (telling the applications software what logic to run when executed), software data input/output control, and the system fault detection-recovery logic.

Given the higher performances in modern embedded and/or real-time computer hardware, the need to create applications-specific operating systems has been greatly reduced. Instead, there are a growing number of general-purpose real-time operating systems available. These can often be tailored to easily accept the code created by the specific computer language used to write the applications software for the Program. Some are even tailored to easily receive applications software created by a specific computer-based software development environment discussed above.

Using a commercially available real-time operating system can greatly reduce the cost and risk of developing a real-time computer-based control system for a program turnaround. Because sizeable design and testing investments are made by the software manufacturer when building these commercially available systems, the capability of the computer system being developed will usually degrade gradually when an error is encountered in the applications software under the control of these operating systems. This response, as opposed to the trauma of just stopping software execution, as has often been the case for the more simple, custom-built operating systems. Using a commercially available operating system also allows for easier movement of a completed software system to a different computer hardware host if needed in the future.

15.10.4 True Polymorphism

Polymorphism was discussed earlier. It is a promise of object-oriented languages such as C++ to allow applications software to *automatically* incorporate common computational functions from a software library. This feature checks that the function selected provides all of the specific performances required by the applications software before incorporating it.

15.10.5 Self-Learning Test Software

As discussed earlier, a major cost in developing software is testing it. The software test designer attempts to test all the functions of the software and provide test-case inputs that test the software to its design limits. Experienced software test engineers are good at achieving these goals. But with modern software logic, this is more complicated than ever, and it has become easier for the test engineer to miss testing one or more executable logic paths in the software.

Software-based systems that test applications software are being developed. The algorithms used in these test systems actively try to find combinations of acceptable test inputs to fail the software being tested. Their intent is to test all the software logic paths in the software being tested, using the entire range of acceptable data and logic values determined during software design.

If these test programs find a test case in which the test comes close to failing, they further vary the test inputs to try to get this particular test to fail. The test software may apply many iterations of different input values in an attempt to find a failing combination. A software test engineer would never have the time to manually derive and apply all these test values.

Software-based test tools provide much more test coverage than traditional manual test methods. As a result, the testing of the target software is more complete. In addition, these software-conducted tests finish in a fraction of the time needed by a software test engineer manually setting up and running the same tests.

15.11 Chapter Highlights

- Exploding breadth of software applications.
  - Growing contributor of product functionality.
  - Often a weak link in the program or project development.
- Software technology—still the "new guy on the block."
  - Man-made science.
  - Capabilities and user rules evolving quickly.
- Software applications are not created equal.
  - Diverse user domains.
  - Application specifics and typical milestones are unique to each domain.
  - Modern software engineers are usually specialists in just in one or a few domains.
- Software is mostly invisible.
  - Cannot be examined like a mechanical device, seating arrangement, electronic circuit, etc.
  - Mostly only has "meaning" when executed.
- Software cost myth.
  - Most software development costs are for requirements identification, design, test, and documentation.

    - Typically less than 10% cost for coding, only 2% to 4% for coding safety-critical embedded software.
- Major savings from correctly designed software.
    - Develops independent software units.
    - Reduces development risks.
    - Numerous other benefits to Program team.
- Unit tests before integrated tests.
    - Low-cost time to find and fix defects.
    - Unit tests must follow a standard and be useable for regression testing.
- What makes an effective software team manager?
    - Software "ambassador" to the development program or project.
    - Must have experience with the domain(s) being developed.
    - Must have experience-based intuition to foresee common development problems.
    - Must be able to completely and understandably describe software milestones and issues to Program leadership.
    - Must verify that the Software Development Plan is complete and is rigorously followed.
- Overkill up front.
    - Make sure Program leadership completely understands the contents and purpose of each software deliverable.
    - Lead the Program software team to stay ahead of planned software delivery dates with margin.
- Software technologies to consider.
    - Automated software development/test environments.
    - Code generators.
    - Off-the-shelf, domain-specific operating systems.
    - Promises of polymorphism.
    - Self-learning software test tools.

Chapter Sixteen

Early Success—"Team Food"

Setting a positive tone with the Turnaround Team must start the first day of executing the Turnaround Plan. As the Turnaround Lead, you and your leadership team have amassed an exceptional team of creative professionals ready to get the Program back on its feet as soon as possible. They are enthusiastic about successfully reaching the Turnaround Commitment on or ahead of its promised completion date. They are prepared to endure some challenging times, and they believe it will be worth it.

So there must not be silence from the Turnaround Lead or Program leadership now. The team is enthusiastically throwing themselves into accomplishing the Turnaround Plan. Most team members appreciate that it will take a while to see the benefits of the Turnaround activity. However, for a leader establishing a new high level of program momentum, the wait time for their feedback to the team on work accomplished with the new plan must be as short as possible!

The Program leadership must transform the team enthusiasm into execution results immediately. Leadership has signed up the team to an ambitious commitment and must now be ambitious about starting to achieve it.

16.1 Let the Team Know of Their Progress from the Start

It is important that the leadership share positive team results as soon as they are available. From the start of executing the new Program Plan, the Turnaround

Lead should conduct a daily status meeting with all program leads, to review the last day's progress, lay out the plan for the next day, share the new risks encountered on the program, and discuss issues that need attention. The Turnaround Lead should consider allowing, if possible, all program hands to attend these meetings at their discretion. It is best if these meetings are conducted in the morning and not last longer than half an hour.

It is important that the leads share any forward progress made by their teams with the team membership promptly, even if the accomplishments seem relatively minor. For example, "preliminary task planning is complete," or "we have recruited eighty percent of our team size target for our task team" are fine descriptions of progress to share. These accomplishments may not be as grand as other pronouncements, but they are still important to the team members—they demonstrate progress.

The implication should be that the new program team is quickly showing its high capability, even if the starting accomplishments are small. Program leadership should especially highlight when an accomplishment is achieved on or ahead of schedule. It is imperative to establish team precedence and the habit of always delivering on or ahead of promised completion dates.

16.2 Leaders Highlight the Power of Teamwork

Program leadership must also identify how communication and support among program team members has led to many of their early successes. Because of the high caliber of the membership of the Turnaround Team, they probably already value teamwork. But it does not hurt to reinforce this behavior with complimentary words describing examples of Program successes resulting from the power of teamwork. Remember, as a leader, you will always be setting the operational tone both by stating what you think is important and by your example.

If one or two weeks go by during the start of the Program Plan without any progress to report to the team, then either the milestones in the Program Plan are not of sufficient detail or the team has stopped making progress. Either cause is serious and must be rectified. The new Program Plan should be designed to a level of detail such that there are at least one or two accomplishments scheduled each week. For larger programs or for any program in which much integration is being performed, there can be over ten accomplishments per week. Again, each accomplishment should be recognized, shared, and celebrated with the program team members.

Whenever program leadership is recognizing team accomplishments, the aspects of the success resulting from teamwork communication should be emphasized. The purpose is to continue to impress upon the team how important fluent communication is among the members of the team as well as between the team members and leadership. There are some successful project managers that have even coined the term "over-communicate" in an attempt to encourage adequate team communication.

It is important for the Turnaround Lead to periodically meet and review with the entire program team the details and progress of achieving the immediate program milestones, progress toward achieving the Turnaround Commitment, and the strengths and weaknesses of the Turnaround Plan. This will be an opportunity for the team to correct any portion of the Turnaround Plan and/or daily work process that they believe can be executed better, and go over the details of the next big program milestone so they maintain sharp focus on its completion. Also, it gives leadership guidance to amend the Turnaround Commitment if a new situation requires it.

16.3 Chapter Highlights

- Sharing Program successes with the team develops enthusiasm to maintain high turnaround tempo.
- Share all successful task completion status with Turnaround Team as soon as available.
- Remind the Turnaround Team they are demonstrating exceptional capability.
- Leaders should take every opportunity to highlight the power of good communication and teamwork the program has demonstrated.
- Program leadership must emphasis and exemplify meeting or beating promised completion dates.
- Program schedules should be of sufficient detail to report completed milestones at least once a week.

Chapter Seventeen

Maintaining Traction

The president of a commercial space company was quoted as saying, "The quickest way to lose money during the development of a commercial satellite is to have an unresolved technical issue." To keep a test failure from devouring program profits, the root cause of the failure must be accurately determined, first try.

I am amazed at how accurately the root cause of a failure can be determined if the right process is rigorously followed. But I am disappointed at how few programs follow good process. It is imperative that a program turnaround precisely executes good root cause and corrective action (RC/CA) process or it will fall behind schedule, exceed cost, and often fail. This process must be thoroughly described in the Turnaround Plan.

If root cause process is not completed, the cause of a fault or failure will probably not be found. The program will end up not eliminating the cause and the fault or failure will occur again, often in an even more integrated and expensive development setting. This reoccurrence will usually be more expensive to determine and mitigate than for the original fault or failure. Also, those who have been in the development program business will agree—the fault will return at the worst possible time!

Accurately determining the root cause of a fault with a rigorous and complete execution of process will cost only slightly more than just applying a cursory process. However, enduring the impact of the original fault again and then executing the cause determination process a second time will cost over twice what it would have if it had been found the first time. This comparison assumes that the second time this fault occurs, it does not result in more catastrophic and therefore more costly consequences than the first time.

Faults and failures that are not completely fixed after the first occurrence usually become a hideous, expensive nightmare. They greatly slow development progress, demoralize the program team, and can cause the project or program to eventually fail.

The dangers of not devoting the resources to find finding root cause is even greater for programs with narrow, fixed profit margins. These programs are often reluctant to exercise the entire root cause determination process because of cost and schedule pressures. Often in this environment, product fixes are quickly applied based on just one or more suspected causes, often a cause that is only suspected is not the real cause. When the fault occurs again, the added cost to follow the complete fault determination process, plus the cost of delaying this program, can cause the program to fail.

Anomalies, faults, and failures are going to occur in any development project or program; this is unavoidable when creating something that has never existed before. When such an event occurs, the successful program promptly finds the real cause, confidently fixes it, and then moves on. I call this successful trait *Program Traction.* A good project or program will continue to use the experiences it has had resolving its faults to better understand the product it is developing. If it has not determined the real cause and implements a correction for a suspected cause hoping it is the real one, it is moving without traction and eventually will be in more serious trouble.

17.1 What Is Root Cause?

Root cause is the starting fault or failure that results in the observed or measured fault. There may be multiple failure symptoms observed as intermediate faults that occurred between the root cause and the observed failure. However, there is inevitably only one root cause.

This is an important trait to check for when a root cause is proposed by the project or program. For example, during the construction of the SR-71 spy plane, basic tests of the strength of the titanium used to build the plane began to fail. Raw material sources, material handling, and material preprocessing were evaluated. Combinations of faults from these were considered, but it turned out the problem was caused by a simple change in the source of municipal water provided by the city of Burbank in California. One root cause.

Another example is of a lightweight structure that was being developed using advanced carbon-based materials. The structure became delaminated under basic vibration testing, and about half of the epoxy-bonded joints were coming apart. Complex fault scenarios combining vibration nodes, poor workmanship, and other factors were being proposed. It turned out that the epoxy would adhere to the carbon sheet material on only one of its sides, but the assemblers did not know this. That was why about half of the joints were failing. Root cause was inadequate review of material specifications by the assembly team. Again, a single root cause (see Figure 17.1).

Faults occur in any kind of project or program regarding any subject. For example, questions that result can include, why are the sixth-grade math scores at the school failing in the last three years? Why does it cost so much to clean and repair newly donated coats to the charity? Why does the onboard computer fail only when the driverless car makes left turns?

As will be discussed later, once it's believed that the root cause has been found, the cause should be interjected during testing to recreate the failure. Of course, this should only be done if recreating the failure this way does not cause substantial delays and cost to the program. Samples of titanium using old and new water were tested. Tests of carbon sheets with epoxy joints on both sides were tested. The results confirmed that the true cause was found.

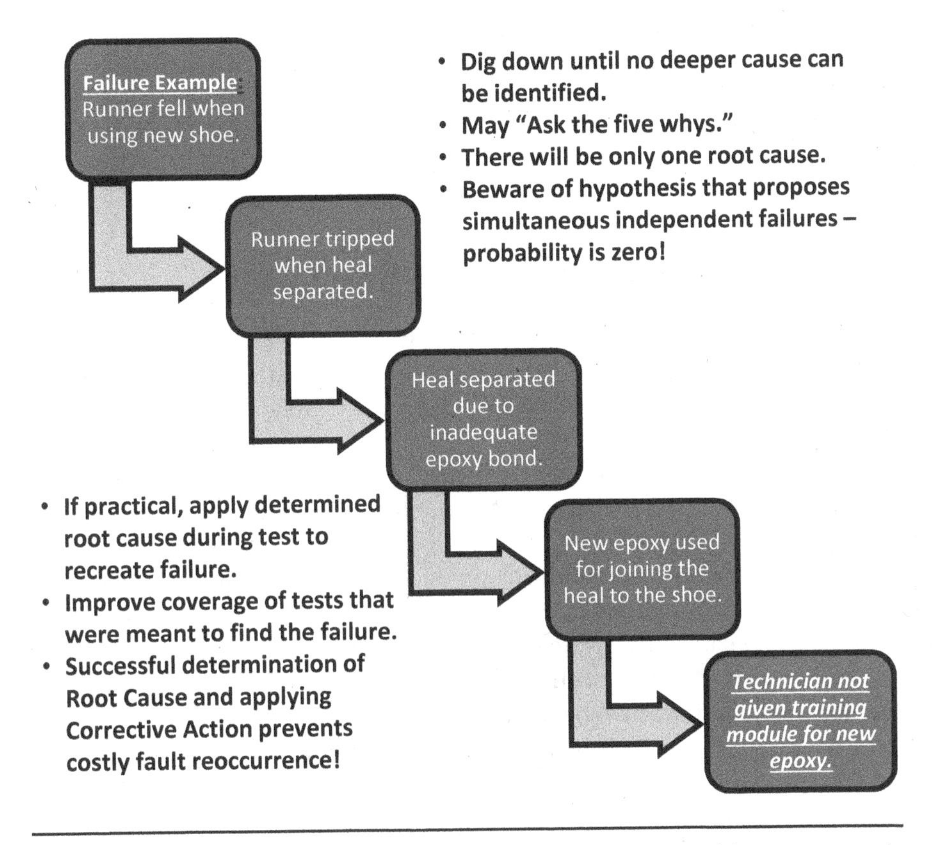

Figure 17.1 The root cause is the originating cause. Be careful not to remedy what is actually an intermediate fault or failure symptom. Otherwise, the fault will occur again—at the worst time!

The fundamental beauty of finding the root cause of a failure and fixing it is that the failure will never occur again, at least due to this cause. This is a highly valuable attribute for any development program!

17.2 Accurate Root Cause/Corrective Action Saves Program Cost and Schedule

I have mentioned how finding and fixing the root cause for a fault the first time will save a very expensive recurrence of the fault. It will usually cost over 100% more to execute the cause determination process again because of a reoccurring failure. Even more cost is incurred as a result of program schedule loss if the failure occurs in a more integrated test environment. There can even be an additional cost to the enterprise from the negative effect on its reputation from having to repeat the cause determination.

A good RC/CA process must be performed promptly. The failure evidence and memories of those who were present at the fault are fresh for a limited time, and this fresh recall is imperative for quickly collecting highly accurate data to determine cause. If the failure review process is delayed, it will often require extra cost to reconstitute the team and data associated with the failure.

17.3 Root Cause Determination

The detailed procedure for finding the root cause of a fault and correcting it will vary between programs and development subject matter, but there are basic steps that should be performed (see Figure 17.2). The following addresses these.

17.3.1 Freeze and Document the Failed Setup

When a fault or failure occurs during product development, the first thing to do is to stop all activity and preserve the test configuration that failed. This may be impossible if what is being tested is active when it fails, so in this case stop the test as soon as practical in a configuration that is safe for the test personnel, the test apparatus, and the item being tested.

Document the test configuration you are preserving. The program must not rely on the memories of the team members to determine the configuration of the test when the failure occurred.

At this point, a preliminary timeline of the events that were performed and observed by the team before and after the fault/failure must be documented.

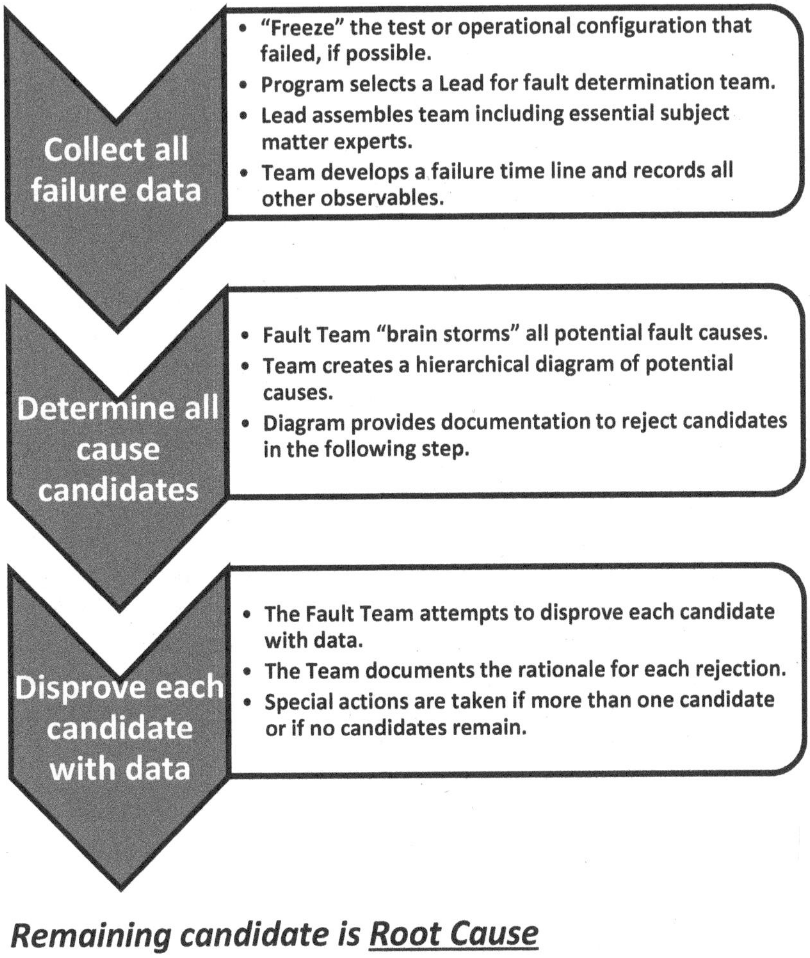

Figure 17.2 Determining root cause accurately and quickly applies a well-known process of elimination. However, it must be followed completely.

Again, this must be performed soon after the fault/failure, while the memories of the test team are still fresh. This timeliness cannot be overemphasized.

17.3.2 Choose a Principle Investigator

Program leadership must appoint someone to organize a team and lead the process of finding root cause and correcting the problem. This person is often called the *Principle Investigator* (PI), and the team may be referred to as a *Failure Review*

Board (FRB). The Principle Investigator does not have to be a senior professional or a subject-matter expert in the problem area; they must, however, have a general understanding of the problem area, the Turnaround Commitment, and the Turnaround Program Plan. They must completely understand the RC/CA process as documented in the Turnaround Plan and be the accountable person to lead it. They must have good planning and leadership skills and have credibility with the program team for being able to organize and lead analytical work.

The Principle Investigator is the single person responsible for leading the work to determine the cause and corrective action for the fault/failure they have been assigned. This person provides all status on the problem solution to the Program. They usually present a final report describing fault/failure cause and correction at a program-level or project-level failure review meeting. This review is mainly conducted to gain approval from the Program for the recommended corrective action. Usually, closing the fault investigation is then contingent on completing the corrective action.

For many programs, a senior management program team including the Turnaround Lead provides this review and permission to perform the recommended corrective action. They make sure the program process for RC/CA was completely followed and the results are well supported by the evidence. They sign off on the final report and data package that documents the determination of root cause and corrective action. The approved final report is usually stored in the Program library or data center.

17.3.3 Assemble the Fault/Failure Investigation Team

It is the responsibility of the PI to select the team members on their FRB (RC/CA team). These members usually include the product development personnel associated with the item that failed, the subcontractor personnel associated with the product development (if applicable), and members from the organization doing the testing.

The PI must have the authority to select who they believe they need on their team, negotiate with program leadership to get the support of these people, give assignments to the individual team members, establish a schedule of regular team meetings to complete the RC/CA process, and make sure the FRB completely follows the process described in the Turnaround Plan.

17.3.4 Complete a Timeline of Events

The first thing the PI must have their team do is to complete the timeline of the events leading up to the failure. Usually the test team will have constructed a preliminary timeline immediately after the fault/failure occurs as part of their initial response to the fault or failure. It is important that this be completed as soon as possible after the failure/fault occurred.

The timeline should include everything that may be relevant to the fault/failure being analyzed, even if the relevance appears to be minor. This timeline will be an important source of data for the analysis to occur later.

17.3.5 Record All Random "Observables"

There may be random observations by team members that could have some relevance to the fault/failure, even though they are not on the timeline. Examples might be, "A transformer for wall power was replaced last year," or "The computer printer software was updated before erroneous reports were printed." These observables should be recorded and used as part of the evidence base to later evaluate candidate causes.

17.3.6 Create Hierarchical Breakdown of Potential Fault Causes

All potential causes of the fault/failure being analyzed should now be recorded by the team. These potential causes are usually the result of brainstorming by the FRB members during a group meeting. They should be recorded in a hierarchical fashion that allows specific candidate fault causes to be listed under fault areas. This can be shown, for example, with a fault tree or an indented listing. An example of part of this breakdown could include:

Fault area	Fault Cause
Workmanship	Poor adhesive bond prep
	Inadequate adhesive used
	Adhesive not cured at correct temperature
Material	Adhesive used was past shelf life
	Raw material surfaces not porous as specified
Etc.	

The purpose is to have the fault/failure team list what they believe is every possible cause of the observed problem in an orderly, easy-to-comprehend format.

Having a large number of fault/failure candidates should not be a concern. Many can be disqualified quickly, but it is very important that every potential failure cause be identified. This is a critical time for the PI leadership. They must make sure the most experienced and imaginative experts associated with the failure determine what all the possible fault causes are.

17.3.7 Attempt to Disprove All Potential Causes with the Data

The FRB now takes all the data they have recorded about the program fault, including test records, the fault/failure timeline, and the fault/failure observables, and tries to disprove each possible cause identified in the prior step. Occasionally when attempting to disprove a cause, more data is needed than had been recorded so far, resulting in the need for more tests to be performed.

Examples of rationale to dismiss a potential fault cause include:

"It was not 'poor adhesive bond preparation' because the same technician has been cementing these bonds for over ten years, receives annual bond refresher training, and has never had a bond fail in the past."

Or,

"Adhesive use-by date was checked and found to be 13 months from expiration. It was found to be stored in an acceptable environment and sealed per the method instructed by the vendor. Therefore, the adhesive used was within its usable shelf life."

Carefully document the argument and supporting data that disproves each candidate fault cause. Accept any data provided from a vendor, subcontractor, or other enterprise organizations as being true. Under the pressure of finding root cause, some FRB members may become suspicious of data provided by sources outside of the enterprise. However, never in my career have I found any organization to be dishonest about the data they provide to a FRB. Most businesses take tremendous pride in what they do and are very aware of the damage that can result to their future if their integrity becomes questioned.

17.3.8 What Is Left?

Now that the team has painstakingly attempted to disprove every potential fault indicated on the fault hierarchy list, there are three possible outcomes:

One Potential Cause Remains That Cannot Be Disproved

Congratulations! This is, with very high confidence, the cause of the fault/failure. This process of elimination almost sounds too easy. Can such a simple process isolate the root cause of a fault/failure so accurately?

Yes, without exception. In the hundreds of fault and failure causes I have witnessed that have been found this way, the same fault/failure never occurring again in what we were developing.

When solved with this process, the fault/failure is always fixed the first time. I have found this to be true after applying this process in many different application domains. Its success is stunning.

When the root cause is determined for a fault/failure, the corrective action is usually obvious. Make sure the corrective action addresses the root cause and not an intermediate failure symptom resulting from the root cause.

Make sure the root cause is really the starting point for the fault/failure that occurred. For example, the cause may be found to be, "adhesive not cured at correct temperature." But the *root* cause may be "replacement adhesive technician had not completed their requisite training before curing the adhesive bond." The corrective action here might be to institute a procedure that assures that any technician working with this adhesive has completed all recommended adhesive applications training within some maximum time span before performing this assembly work.

More Than One Potential Cause Remains That Cannot Be Disproved

Occasionally, after attempting to disprove every cause candidate, more than one is left. When this happens, the team should review all test results, including test data, the timeline for the fault/failure, and the other test observables, to check again that all the data was considered when eliminating the fault candidates and that none of the remaining fault candidates can be disproved.

If this unusual event does occur and a single cause cannot be isolated, then complete corrective action must be applied to each remaining fault. This is often referred to as "worst case" corrective action.

The result of this type of corrective action is as successful as when the single cause of the fault/failure is known. However, it has two disadvantages. First, it costs more to correct multiple root causes; and second, the program will not know which of these remaining root cause candidates was the *real* root cause, making it more expensive to avoid this cause in the future and diluting the Program's understanding of their product.

All Potential Root Causes Are Disproved by the Data

In the hundreds of root cause investigations I have led or approved, this has never happened, but it can occur. If it does, it indicates one of three problems:

- Some of the failure data is in error, and this data has caused one of the candidate fault/failure causes to be wrongly disproven. This candidate may in fact be the cause of the observed fault/failure. If this is suspected, the failure data should be reviewed for correctness. Possibly the test may have to be run again.
- The FRB did not identify all the potential causes of the observed fault/failure. If they had, one would have been isolated as root cause by the test data. If this is suspected, the fault team must review the test data, timeline, and observables and try to brainstorm additional potential causes that may have been missed.
- The measurements originally indicating there had been a fault/failure in the product during test are in error. The test data cannot isolate a root cause because there had never been a fault/failure. Instead, there could have been an error in data analysis or a problem in the test setup that gave the appearance of a fault/failure. The test setup, data observed, and process applied during testing must be reviewed to verify that the fault or failure actually occurred.

In very rare instances, the root cause is not found after the actions taken in the above three bullets. This results from not gathering all failure data when the failure occurred. Often, from not following process. The unresolved cause of this failure will place future Program work at some risk. This is avoided with strict process-based response to all faults or failures.

17.3.9 Recreate the Fault/Failure to Cinch the Deal

Within practical limits, recreate the failure and observables by inducing the root cause that has been identified. Showing that this cause produces the fault/failure with the other documented observables is the last action necessary to verify root cause.

Sometimes the nature of the failure could damage valuable product or test equipment, and recreating it may not be worth this risk. In this case, sometimes either the failure can be recreated with simulations or analysis, or it can be allowed to propagate through only some portion of the product without reaching parts that might be damaged.

Starting corrective action without completely demonstrating that the derived root cause results in the observed fault/failure is a low-risk, but sometimes necessary, alternative.

17.3.10 Again, Accept Only One Root Cause

I would like to offer a few suggestions to verify that the root cause the Failure Review team has determined is the correct one.

Only accept *one* root cause. If a team is proposing a compound of faults that have caused the fault/failure, they are wrong. The one real cause may have led to multiple fault steps between the root cause and the observable or measured manifestation of the fault, and this can be where the confusion lies. The FRB may be analyzing just the symptoms of these intermediate steps. If so, they must drill deeper to find the single root cause.

Some schools of RC/CA analysis recommend that the investigators ask the "five whys." The purpose is to keep asking why a fault/failure symptom is observed until there is no deeper answer to this question. At this point, you are presumably at the originating, or root, cause.

One of my tests for finding the single root cause is being able to describe a single action that will fix it. For example, "The printers in the tax service keep jamming because the paper is too thin." The correction is to use the thicker paper that is approved by the printer manufacturer. Or, "The thin film capacitor in the radar electronics failed because inadequate staking material was applied during construction." The correction is to establish that new technicians must complete a special training class before they can apply this staking material.

Finally, let me bring up an erroneous root cause that I have often encountered. It occurs when electronics are used in the product that had the fault/failure. The error is that a cause is proposed that requires multiple specific failures by certain components in an electronic circuit. Maybe this type of cause is postulated more often because of the increasing amounts of electronics used in modern products.

However, please recall the edict stated earlier. Ultimately, there is always one root cause. The chance of one of the components in the circuit failing, as suggested, is usually very small. The chance of more than one component failing in the way suggested at the same time is essentially zero! Unfortunately, it is usually easy to predict a wide range of fault/failure symptoms by choosing some combination of failures in the electronic components in a product.

One example of this was a report from a subcontractor's failure investigation team working for a program I was leading regarding slightly skewed measurement values from an instrument they were providing. This subcontractor had an outstanding reputation, which was well deserved. But they provided a root cause for this anomaly that stated that four components in their analog electronics failed simultaneously. Their corrective action was to perform more screening/testing of these components in the future. They recommended a workaround for the instruments already assembled.

The report was timely, complete, and very professional—but I refused to accept it.

A vice president from the subcontractor called me. He was very frustrated. He knew the performance incentives for them might be in question. He was convinced that these four electronic components were the cause. I told him that unless there was some common link in the manufacturing or their application that might cause them all to fail together, the chance of all four failing at the same time was virtually zero. I asked him to please continue his failure investigation.

Three days later, I found a folder on my desk from this subcontractor containing photographs of the root cause—it was contaminated optics.

17.4 Same RC/CA Processes for All Turnaround Program Team Elements

For a program turnaround to succeed, all members of the program that are developing and providing the product—including the program's subcontractors and suppliers—must have a strict RC/CA process. These sources should be required (by their contract or purchase agreement if applicable) to immediately communicate any acceptance test failure for material they are delivering. They should then provide a report on the cause of the failure and how they corrected it. Sometimes their RC/CA team will include members from the Turnaround Program.

The least error-prone and most effective communication of development failure status and closure status will occur between the Program, vendors, and subcontractors if they adhere to the same RC/CA process. It is highly advised that this requirement be included in their contracts.

17.5 Chapter Highlights

- What is root cause?
  - The starting point of a fault or failure.
  - Ask the "five whys."
  - If root cause is corrected, the failure will not occur again.

    - If possible, derived root cause should be applied to recreate the failure to verify it.
    - Root causes conjectured to be the result of multiple simultaneous failures are virtually impossible.
- Accurately determining RC/CA with process is imperative for Turnaround success.
    - Prevents very costly fault reoccurrence.
    - Following a well-documented process is successful much faster than an ad hoc response, saving cost and time.
- Example root-cause determination procedure.
    - Freeze the failed configuration if possible.
    - Assign a Principle Investigator.
    - Select a fault investigation team.
    - Create a timeline.
    - Record all associated "observables."
    - Create a hierarchical list of potential fault causes.
    - Attempt to disprove each potential cause with data.
    - Fault remaining is root cause.
    - If more than one candidate remains,
        - Not all fault-related data was recorded.
        - May require a "worst case" correction.
    - If no candidate remains,
        - Not all potential causes were identified.
        - Error in the data recorded.
        - The suspected fault/failure did not actually occur.
- Completing RC/CA successfully can be in danger in highly schedule/cost-driven programs.
- The program or project RC/CA process must be required in subcontracts and vendor agreements.
    - Assures adequate correction of faults at the supplier.
    - Common process supports accurate communication of RC/CA status between program and suppliers.

Chapter Eighteen

Shackle the Configuration

There is a saying in airplane development that applies to all development programs today: "Test as you fly, fly as you test." When verifying/validating a new process or design with test, the content of this process or design must then remain unchanged when it is implemented.

I appreciate that this may sound obvious to the reader, but it is amazing how frequently a slight change to the configuration of a product after it is tested causes a product to fail when it is placed into operation.

A mindset must be established in the Program that even the slightest addition, deletion, or change to process or design can result in an unanticipated interaction in the product that causes a failure.

18.1 Examples of Errors with "Test as You Use"

- A grammar school is developing a new math curriculum in an effort to improve the state's math test scores for tenth graders. A computer database is developed to track these scores in three districts so they can be used to improve the curriculum.

  This new math curriculum is now being implemented and tested in one of the districts. The teaching staff elects to use a different computer database that only recognizes that one district. The rationale is, why include database scope for the other two districts if they are not a part of this first trial.

  Yet, there may be some undiscovered interaction among the databases when all three are integrated. If the other two databases are not included in this trial, a database error may not be discovered until the full deployment of the new math curriculum to all three districts.

- A company is building a self-flying airplane. It will require a pilot to be on board if passengers are on board. The pilot will monitor the instruments and take over control if the automated functions fail.

  The self-flying capability is being test flown with no one on board. In the interest of saving money, the test director decides to fly without cockpit instruments on board. The rationale is that there will be no one on board to use these instruments.

  However, it is vital to realize that these instruments may affect the operation of the rest of the self-flying capability in ways no one could predict. Asking the most informed experts on the system if the addition of these cockpit instruments will affect its performance is asking them to prove a negative. The best they can assert is that "it should not." Yet when tested, it is often discovered that a portion of the process or system affects another part of it in ways no one had imagined. These unanticipated interactions are one of the reasons we verify and test a product to begin with.

  For this example, when being tested, the unmanned aircraft must have all the flight instruments on board, connected, and operating in the final configuration in which this aircraft is intended to be used. This, even if there will be no one on board to look at these instruments during this test.

18.2 Common Program Configuration Mistakes

As you may see, any change in the configuration of what is being developed can affect its function. It is usually impossible to predict what side effects will occur, even for simple products. To stay within the cost and schedule constraints of the Turnaround Program, we must test in exactly the same configuration in which we will apply the new design or process. No configuration change can be allowed unless we thoroughly test it again.

In the following, I highlight a few common ways dangerous configuration changes can occur without the Turnaround leadership knowing it (see Figure 18.1).

18.2.1 Parts Substitution

The Turnaround Program must be aware of any part in the process or product that is changed or replaced, even with a so-called "equivalent part." This

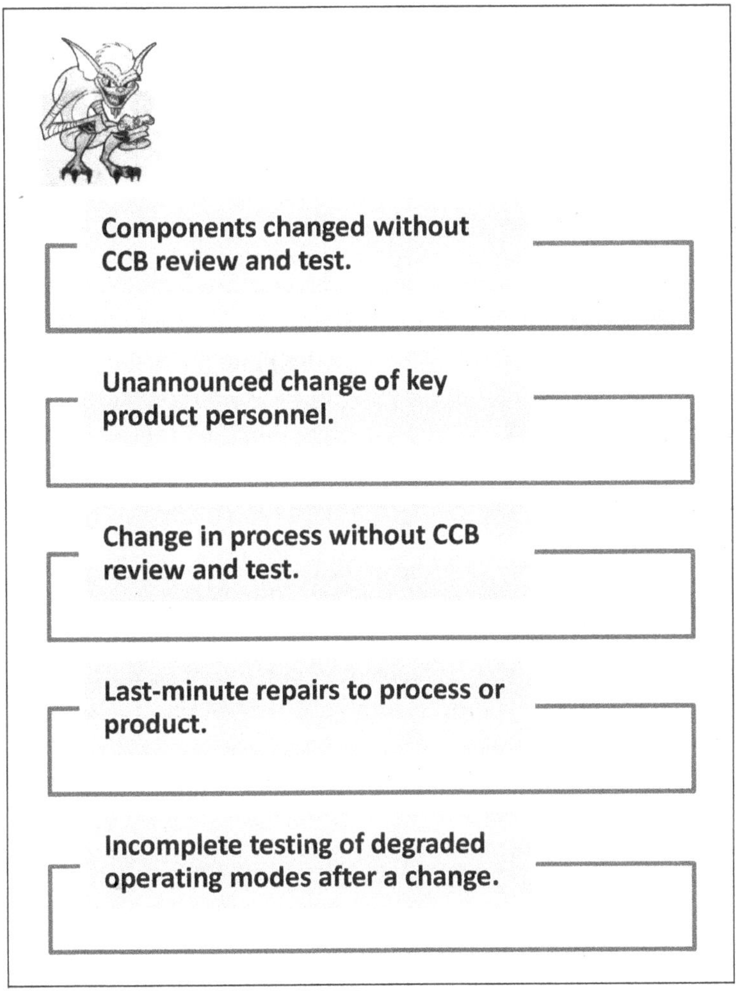

Components changed without CCB review and test.

Unannounced change of key product personnel.

Change in process without CCB review and test.

Last-minute repairs to process or product.

Incomplete testing of degraded operating modes after a change.

Figure 18.1 These are some of the configuration gremlins that can ruin any product. They often occur frequently as the product is nearing completion. All project or program participants must maintain their committed process of thorough configuration control throughout the program timeline.

includes all parts used and provided by suppliers and subcontractors. It should be stipulated in the Turnaround Plan and in all subcontracts and agreements with program or project suppliers that any change of a part under any condition must be approved by the program's Change Control Board (CCB). If the need to change a part is urgent, an Emergency CCB can be called.

The CCB must verify that the need for the replacement part is merited and that the performance of the replacement part, based on inspection and/or analysis, will not change the performance of the product it is used in. The CCB must determine how much regression testing will be needed to fully validate the product after the part is changed. Usually, regression testing will be necessary, even if the part being replaced performs a very simple function.

Even products provided by a subcontractor or supplier that contains parts that have not been changed can vary in performance over time. This is again why a successful program will continuously monitor for acceptable test results and trends in metrics provided from all subcontractors and suppliers.

Changing the source of a part used by the Program, a subcontractor, or a supplier, even if it is claimed to have the exact same form, fit, function, and part number, is a major event and must be approved and introduced via CCB direction. Again, regression testing will probably be necessary. In addition, the record of the part change will be important product history if there is a product fault or failure in the future.

For a program turnaround to be on step, it is essential that the program write into every subcontract and purchase agreement that any change in a part or source of a part must be approved by the Program CCB. This may seem excessive, but it will save valuable program expense and time. Penalties for not complying with this process should be written into the subcontract or purchase agreement.

18.2.2 Changes in Personnel

If, for example, the "Leader of Promotions" retires from a nonprofit organization that is the subject of a Turnaround, the replacement for this major role should be checked by the Program CCB. The background of the replacement should be reviewed to make sure they have sufficiently experience to provide the same services as their predecessor.

If a subcontractor or other supplier is providing a product for which some critical assembly steps require workers with special training and/or other special talents, the CCB should be informed if there is a plan to replace one of these workers. Consider having the supplier list by name the experts critically needed to provide their product.

The subcontract or supplier agreement should include that they must report any intended change in key personal within some specified time before it is made. The purpose is not for the prime contractor to manage the suppliers, but the Turnaround Program must be aware if any critical member of the team is being replaced. Any approval required by the Program for changes like this should be stipulated in the subcontracts and purchase orders. The process for these approvals must be explained in the Turnaround Plan.

18.2.3 Changes in Process

All changes in the processes described in the Turnaround Plan or in any of the processes used by the suppliers and subcontractors must be reviewed and approved by the CCB. These can include changes in the processes specified and used in all procedures, assembly steps, test plans, plus changes in processes due to any changes in organization and reporting structure, and more—in fact, any process at all.

If, for example, the way expenses are being accrued is changed for a new business approach for a taxi service that had to initiate a turnaround, this process change must be reviewed and approved by the CCB.

18.2.4 Last-Minute Changes

It is human nature to wait until the last minute to apply final changes, if for no other reason than to make sure the team has identified all of them so they can make them at one time. Therefore, often just before a product delivery milestone, there are a large number of last-minute changes that are proposed/implemented. In the rush to complete under the pressures of achieving a delivery milestone, the Program may not be aware of some of these changes without specific provisions in the subcontracts, purchase agreements, and Turnaround Plan to catch them.

For example, for electronic boxes, there are sometimes last-minute circuit wiring changes to improve performance or fix a fault. For a new water treatment plant, there may be an added filtration step, "just to make sure," right before the new treatment process is implemented.

Although these last-minute enhancements are added with the best intentions, they often turn out to be very dangerous, simply because these changes may not have received the high level of test and verification that had been given to the rest of the system.

Without extensive testing and verification, these last-minute changes can become a very serious problem source that will ruin the successful deployment of the product. The Turnaround Lead and the entire program team must be vigilant of these changes and not allow them without the same planned, thorough test and verification the rest of the product has gone through. Usually, this will include a review and approval by the CCB of the details of the proposed change and needed regression testing.

It may require tremendous patience from the Program, but leadership cannot allow any last-minute changes to be made in the product, from any Turnaround Team member, without the Program being informed of the change and authorizing it.

18.2.5 Change of Parts Source

Even if two sources of raw material or piece parts for the Program, program subcontractor, or other supplier are said to adhere to the same industrial specifications, there are often subtle differences between them.

The Turnaround Program must be informed when the Program, a subcontractor, or a supplier changes a source of piece parts or raw material. This change must be reviewed and approved by the Program's CCB. This action will likely catch any subtle change that may affect the performance of the product they are creating. In addition, this action will document this change that may be a candidate for root cause if a failure occurred after the change was made.

18.2.6 Poorly Tested Degraded Operation

For automated products, usually much more time and money are spent making the system successfully recover from erroneous inputs or other system faults than are spent making the system perform its basic function. Program leadership should be especially aware that after a product configuration change, the fault recovery functions must also be completely tested again.

It is easy for fault recovery to not be completely tested during regression testing. This is because the fault recovery system is usually considered to be a backup function. Of course, during the completion of rigorous development schedules, the major emphasis will tend to be on completing the primary product function.

Program leadership must insist that fault recovery functions receive the same high level of test and verification that the rest of the product receives, both during initial product testing and after changes are made to the product.

18.3 Subcontractors and Other Suppliers Must Follow the Rules

This problem of a subcontractor or supplier making a change that falls under a Program's radar can occur more often than one might think.

For example, the Program may not be informed that the part is being changed in a product being delivered to them, either because the supplier does not believe the change is significant or because the need to inform the program was not written into the contract or purchase agreement. Sometime later, there is a fault/failure in the Program's product and root cause is eventually traced to the supplier's replacement part.

This kind of problem usually takes longer and requires more cost to find and correct because it requires a hybrid of program and subcontractor/supplier personnel to finally drill down to the cause.

The supplier or subcontractor will often feel defensive about taking blame for this kind of problem. They may rightfully claim that the need to inform the Program of the parts change was not written into their contract. They may even say that it is their business to make what they consider to be minor parts changes if they have determined it will not change the performance of what they are delivering.

The need for the Program subcontractors and suppliers to freeze and control the configuration of what they provide is just as important as doing so within the main Program.

It is imperative for a Program Turnaround that the subcontractors and vendors follow the same rules for maintaining their product configuration as the Program does. These rules must be written into the subcontracts and purchase agreements.

Applying a system of awards and penalties in the subcontract and supplier agreements to motivate strict adherence to configuration control guidance should be considered.

18.4 Periodic Quality Metrics

As discussed earlier, the subcontractors and other suppliers must maintain a record of their quality metrics. Provisions should be established that allow the Program to examine these records at any time.

These metrics will document that what is delivered meets the requirements of acceptance testing. Just as important, this data will reveal any trends in the acceptance test measurements. If these trends show movement toward unacceptable values, the Program can often predict when the products being delivered by the supplier will fail acceptance testing and can start remedial action before product failures occur.

Stable and acceptable values for supplier metrics is another way of being sure the products provided by the Program's suppliers are maintaining an unaltered configuration.

18.5 Program Change Board Moderates Changes

As mentioned earlier, the Turnaround Program's CCB will often consist of senior professionals and subject-matter experts from the Program, possibly as well as experts from the customer and subcontractors. This board will typically allow fewer changes to the processes and products as the Program completes final validation/testing of what it is making. This helps in motivating the achievement of the final product configuration within the time necessary to achieve the Turnaround Commitment.

18.6 Chapter Highlights

- Test the product or system exactly as it is used, use it exactly as tested.
- Common configuration errors leading to failure include:
    - Parts substitution with insufficient documentation and test.
    - Changes in process with insufficient documentation and test.
    - Last-minute repairs.
    - Incomplete testing after a change.
    - May occur at a supplier or subcontractor.
- Subcontractors and suppliers must adhere to their configuration control plan.
    - Report all test failures, parts changes, process changes, supplier changes, key personnel changes, and quality metrics within a maximum time span.
    - Reduces turnaround cost if their process is similar to the Program's process.
- CCB can motivate Program process and product work to complete.

Chapter Nineteen

Document and Follow

Clarence (Kelly) Johnson remains a major inspiration for those striving to take a small team, get them on step, and develop a highly innovative system. He led the famous Lockheed "Skunk Works," which developed the first U.S. production jet fighter, the T-33. He also led the development of the U-2 and SR-71 reconnaissance airplanes, as well as many other aircraft. He was noted for assembling a small number of professionals, with all the specialties needed to develop a breakthrough airplane located in one environment (colocation), and leading the daily activities to achieve a single, often difficult, but easy-to-describe goal (Turnaround Commitment).

Kelly Johnson was a master of the efficient use of time and resources. For example, he led the development of the T-33 in just 143 days with a team of only 140 professionals. They started by working together in a large rented circus tent. Any team member had to walk only a short distance to communicate with any other team member. Communication was conducted face to face.

Throughout the life of the Skunk Works, the employees had a well-deserved pride. Some felt they could build any airplane they could imagine. They were a small team and highly effective. Their leader kept their commitments pinpoint simple. They were on step.

In Kelly Johnson's book, *More Than My Share of It All,*[1] he reviews his fourteen operating rules. One of them is, "There must be a minimum number of written reports required, but *important* work must be recorded thoroughly."

---

[1] Johnson, Clarence L. "Kelly" (1989). *More Than My Share of It All.* Washington, DC: Smithsonian Books.

I will review some of the documents that are important for a program turnaround. The benefits of generating these documents are far greater than the investment to create them. They will help the program team to achieve the Turnaround Commitment as quickly as possible with low risk.

19.1 Turnaround Plan (New Program Plan)

This document is a detailed description of how the Program will be executed. A copy must be given to every team member.

A good measure of the value of the Program Plan is how much it is used by each team member. The more its pages become frayed from use by the team members, the better. Whenever there are questions or disagreements among team members about the operation of the Program, they should reach for their copy of the Turnaround Plan. It will often also act as a table of contents to the other documents that support the Program, such as the Software Development Plan and the Risk Management Plan.

The Program Plan should be easy to read—the more succinct the better. It is usually written by the Turnaround Lead and his or her leadership team. It must be approved by the customer and enterprise management.

Some details for the Turnaround Plan are often still being written while the Program is starting. This plan may be considered a first vision of subsequent Program Plan revisions. The outline for the Turnaround Program Plan was shown in Figure 2.3 and is repeated here as Figure 19.1 for convenience.

19.2 Program Requirements Document

This document contains all the program requirements from the customer and the enterprise. It may provide an overview of the requirements flow down to the Program tasks, the subcontractors, and other suppliers.

- **Executive Summary**
  - **Turnaround Commitment statement**
  - **Description of how to reach commitment**
- **Organization**
  - **Organization Chart**
  - **Role and responsibilities**
  - **Change management**
  - **Facilities**
  - **Project completion**

- Subcontractors and vendors
- Rules and expectations

- **Scope Management**
  - **Scope Summary**
  - **Requirements Management**
  - **Configuration Management**
  - **Deliverables**
- **Schedules**
  - **Master Schedule**
  - **Schedule Control**
- **Cost**
  - **Estimation Process**
  - **Budget Allocation**
  - **Budget Control**
- **Quality**
  - **Monitoring**
  - **Control**
- **Human Resources**
  - **Acquisition**
  - **Development/mentoring**
- **Program Interfaces**
  - **Stakeholders**
  - **Reporting and Communication**
  - **Team interdependencies**
  - **Metrics Collection**
- **Risk Management**
- **Procurement**
  - **Subcontractors**
  - **Vendors and services**
- **Program Information Management**
- **References (potential)**
  - **Integration and Test Plan**
  - **Quality Plan**
  - **Safety Plan**
  - **Headcount Plan**
  - **Product Support Plan**
  - **Software Development Plan**
  - **Risk Management Plan**
  - ***more***

Figure 19.1 This is a typical outline used for constructing the Turnaround Plan (a Program Plan). As discussed in Chapter 2, this plan describes how to achieve the Turnaround Commitment and complete the Program. It is the primary source of direction for the Program and is under configuration control.

The customer requirements most often include all required product performances, content of requested deliveries, delivery dates, special team rules and protocol during development, cost constraints, reporting contacts and protocol, customer terms and conditions, etc. If the work to be done is sizeable and expensive, some customers will convey these requirements with a combination of a Statement of Work document, a Requirements Specification document, a Work Breakdown Document, a Funding Profile document, a Master Schedule, and often more.

Any program team member should be able to access these program requirements and contracts unless this access is constrained by sensitive data or customer request. This material simply provides the basic completion requirements for the work every team member is trying so hard to accomplish and should be accessible if possible.

19.3 Integrated Schedule

A good Integrated Schedule provides the detailed schedules that support the Program's Master Schedule. It also shows the interdependencies of each task. Figure 19.2 shows an example of an Integrated Schedule using a popular computer-based program planning tool.

No task must be scheduled to be performed at a given time before all tasks providing required input have done so. For complex programs, this schedule should be maintained with a computer-based scheduling system. This will greatly reduce the scheduled maintenance workload and errors that are more likely to occur with manual scheduling.

Usually, each program task team or department will maintain their individual detailed schedules. However, their need dates and delivery dates must feed up to the Integrated Schedule. Updates to these interdependencies should be reviewed, approved by Program leadership, and incorporated at least weekly.

The Turnaround Lead must have a clear policy for schedule margins. If the Lead is incorporating multiple schedule "books" to manage the Program, there may be amounts of schedule margin known only to a subset of the Program leadership.

Part of the Lead's schedule policy should describe the approach for adding margins into the schedules of the individual task teams. The task leads must be responsible for deriving this margin.

These schedule margins at the task level should be identified in the Integrated Schedule. As mentioned before, when a task team slips a commitment date in the Integrated Schedule, it should be considered a serious infraction. The schedule margins are designed to absorb delays resulting from development problems.

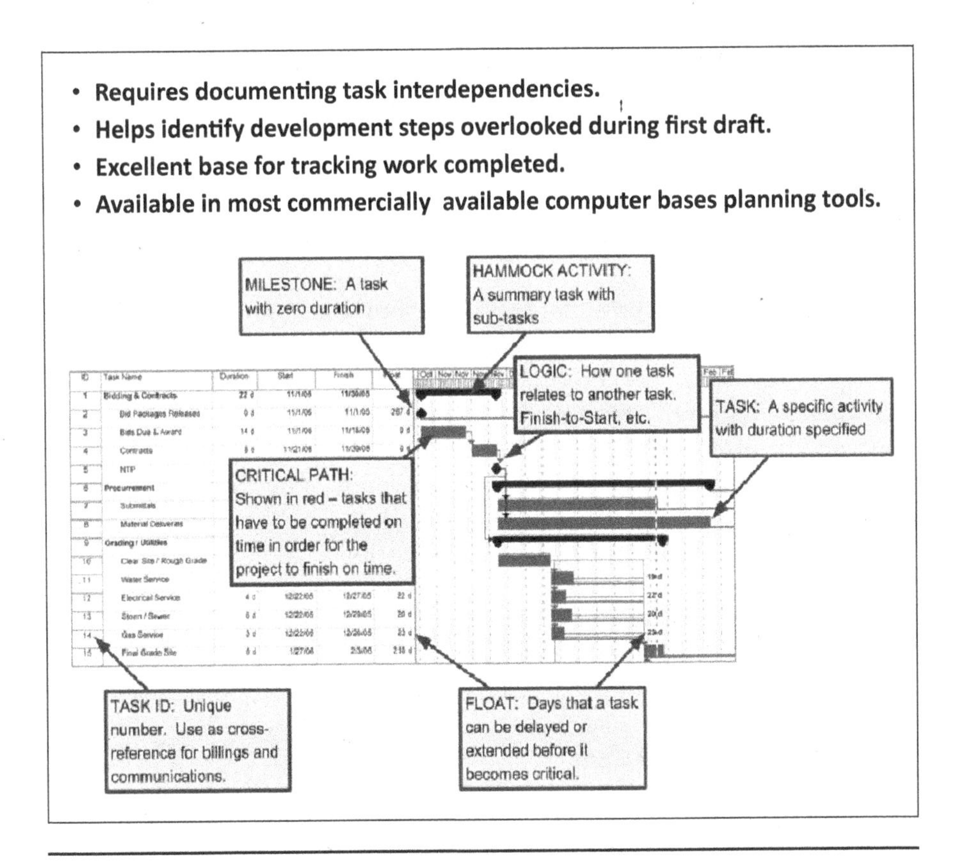

Figure 19.2 This is an example of the output of a computer-based tool for maintaining the Program's Integrated Schedule. It will identify the critical path and provide reports in any of a number of conventional schedule formats.

This margin must be allocated and monitored as the only schedule buffer for these problems. Delays beyond the margin must be evaluated as a planning error by the task manager or lead. As discussed earlier in this book, the task team must still attempt to achieve the promised completion date with extra work effort.

Subcontractor and supplier schedules must clearly flow into the Integrated Schedule. These organizations must also be held strictly to their commitment dates. All subcontractors and suppliers must be required to establish a system of schedule margins that ensure that they achieve their committed dates for delivery.

The Turnaround Lead should consider applying a system of early delivery incentive awards and late delivery payment penalties for some supplier deliveries. The use of these incentives may increase the total program cost but are highly advised for tasks that are especially critical to achieving the Turnaround Commitment on time.

The Integrated Schedule for the Turnaround Program should be updated at least weekly. A current version of it should be easily accessible by everyone on the program.

19.4 Risk Management Plan

With the passion and optimism associated with a new program or program turnaround, a Risk Management Plan is often overlooked. Yet it is just as important to manage risk throughout the life of a Turnaround Program as it is to achieve cost limits, schedule dates, product performances, and other customer requirements.

A risk management matrix is usually developed to determine the severity of each risk and its impact if it becomes as issue, as shown in Figure 19.3. Figure 19.4 shows an example of the detailed documentation used to record the rationalizations used to compute these severities.

The Risk Management Plan should specify periodic reviews by a Risk Management Team throughout the life of the Program. The Risk Management Team must identify the risks to meeting the Turnaround Commitment and assemble risk abatement and issue contingency plans to remove risks and respond if risks become issues.

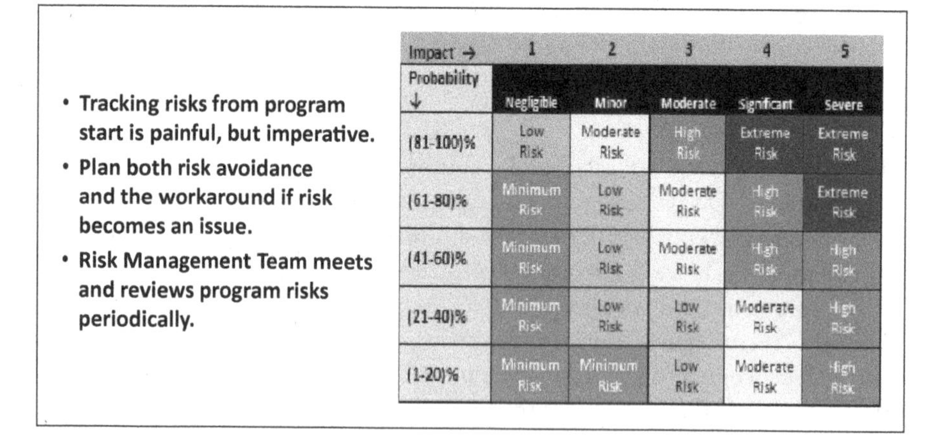

Impact →	1	2	3	4	5
Probability ↓	Negligible	Minor	Moderate	Significant	Severe
(81-100)%	Low Risk	Moderate Risk	High Risk	Extreme Risk	Extreme Risk
(61-80)%	Minimum Risk	Low Risk	Moderate Risk	High Risk	Extreme Risk
(41-60)%	Minimum Risk	Low Risk	Moderate Risk	High Risk	High Risk
(21-40)%	Minimum Risk	Low Risk	Low Risk	Moderate Risk	High Risk
(1-20)%	Minimum Risk	Minimum Risk	Low Risk	Moderate Risk	High Risk

Figure 19.3 This type of matrix is the heart of most modern Risk Management Plans. It is an easy-to-comprehend visual aid that inventories program risks and their likelihood of becoming issues. Its application is necessary to reduce surprises and plan contingency responses during a Program Turnaround.

#	Risk (Hold Mouse over every cell in this row for 'Comments')	Outcome	Existing risk treatment actions in place	Likelihood	Consequence	Rating	Proposed risk treatment actions to mitigate risks (to reduce level of risk rating)	Additional resources	Target date & responsible person
1	Bushwalking/rafting in winter conditions - Participants suffers from hypothermia	Serious possibly fatal Consequences Involvement of emergency services Possibility of cancellation of similar events in the future Coronial or Ministerial inquiry, 'bad press'	Participating school to complete	3	4	H	Pre-excursion briefing of participants, ensuring appropriate clothing/equipment is brought along Frequently check for signs of hypothermia during the activity, especially in cold/wet weather Leader/key participants to provide emergency equipment (e.g. tent, space blankets, fire lighting equipment, first aid kit etc.)	Participating school to complete	Participating school to complete
2	Surfing - Participant swept out in rip or dangerous current	Potential of injury/death to participant and intended rescuers Involvement of emergency services Possibility of cancellation of similar events in the future Coronial or Ministerial inquiry	Participating school to complete	4	2	M	Pre-excursion briefing of participants Ensure only competent swimmers participate Adult observer keeping watch Rescue/communication equipment available Appropriate qualifications for excursion leader	Participating school to complete	Participating school to complete
3	A medical emergency will arise (pre-existing condition)	Injury/death of student Curtailment of excursion Future excursions of that type put in jeopardy Additional costs to student and	Participating school	3	2	M	Double check to make sure all relevant information is available Remind student/participants prior to the excursion of the importance of taking along medication & ask person again on the day Take along school issued	Participating school	Participating school

Figure 19.4 This is an example of detailed risk analysis documentation that is used to populate the risk matrix.

As mentioned, each identified risk is usually rated for probability of occurrence and impact of occurrence. A merit of importance is then computed based on some combination of these two values. These merits of importance are used to establish a hierarchy of risk importance for the Program. Usually, the potential issues with the highest risk and greatest impact will get the most attention.

This risk process is necessary to keep the Turnaround Program from suddenly stopping and becoming lost in bewilderment when an issue occurs. This process forces the Program to address and solve problems about which many team members may be worried but are reluctant to acknowledge.

Care must be taken to clearly delineate the high-impact risks to the Program from those that are of lower importance. Otherwise, the risk management process can become overwhelmed with risk identification and abatement planning for risks that have a very low probability of occurring and/or have little or no impact.

As mentioned, the Risk Management Team usually meets once a week. The composition, function, agenda, and authority of this team should be described in the Turnaround Plan and the Risk Management Plan. As a minimum, during these meetings, the Risk Management Team will report on current risk and issues status, review new analysis, and record and evaluate any new risk concerns and issues.

19.5 Expenditure Profile Plan

The Expenditure Profile Plan shows the allocation of funds to the program elements by the Turnaround Lead. It usually divides the allocation into monthly or weekly divisions depending on the size of the Program. For some programs, this plan may show the flow down of funds to the individual Program Task Teams/departments and overhead functions, if it is of value to the Turnaround to track expenditures at this level.

Management reserves and other funding margins held only by turnaround leadership are usually not shown in this plan.

The Expenditure Profile Plan describes the allocation of funds with which the Turnaround Commitment must be completed. If a program team spends more than their allocated funds, the amount of this overrun is usually taken from the allocation to the other teams. Otherwise, a cost overrun for a Turnaround Program may quickly jeopardize its credibility. Many customers can neither afford nor tolerate a cost overrun. A Program-level cost overrun can be very detrimental to successfully completing a Turnaround Program.

This funding plan may also identify the funds that will be allocated for future work after the Turnaround Commitment is achieved.

19.6 Software Development Plan

Software is providing an increasingly larger portion of the functionality of many new systems being developed. If a program turnaround is developing a product that includes new software, a Software Development Plan (SDP) is essential.

Why a special plan for developing software? Because as discussed earlier, software is less tangible, has many parts, and is a new and rapidly evolving technology.

Most of the Program Turnaround Team has only a cursory understanding of the development of software for the Program. The Software Development Plan will describe in detail the software development methodology, the development tools used, and the software development milestones. It will also describe in detail the development steps of the software, including software requirements flow down, software design, coding, peer reviews, unit test, integrated test, and support for product integration. It will highlight the process and rationale for estimates, including the computer memory sizes needed, the number of software instructions per second needed to execute the software tasks, the computer hardware resources needed, and the durations for the software development schedule.

The SDP serves two primary audiences. First, it is the plan that guides the work of all the program software developers. Each developer must be very familiar with the SDP and follow it. When there is any disagreement between

Turnaround Team members about the software development approach or development flow, the first thing they should address is the SDP, which will provide in-depth technical details when specifying the different portions of the development plan.

Second, the SDP will provide an overview description of the software development process for all the non-software program team members. For example, if a preliminary design review on test software is scheduled, the SDP will describe what minimally must be conducted at this review and how the results will be evaluated. Program Team members, enterprise leaders, and customer team members should be able to understand this information from the description in the SDP. As discussed earlier in this book, the well-chosen Software Manager will answer further questions about details of the software development work.

The SDP may seem unnecessary in the minds of some members of a program turnaround team. Unfortunately, some development programs have come to this conclusion. Some programs have evaluated the SDP as a nice thing to record, but their software team will end up "doing it their own way." This kind of approach will often seriously hamper or destroy a program turnaround.

The SDP should contain the development guidance and laws for all the software being developed by the Program Turnaround. If it is flawed, the change process should be used to revise this document, usually by the program Change Control Board.

The best SDPs are short and simple to comprehend. Every member of the Program's Software Development Team and Program Team leadership should have a copy of it.

19.7 What Is Necessary?

There is often a fine line between documenting information on a program that may make the participants feel safer but brings little value, and documenting information that is key for program success. It is essential for the Turnaround Program leadership to make this distinction and to exclude unnecessary documentation throughout the life of the Program. The only exception is any special documentation that is required by the enterprise and/or customer.

19.8 Chapter Highlights

- ". . . All *important* work must be recorded thoroughly!"
- Important Program documents include:
  - Program Plan.

- Integrated Master Schedule.
- Risk Management Plan.
- Expenditure Profile.
- Software Development Plan (if applicable).
- Additional documents that are often needed:
  - Program or Project Contracts to customer.
  - Subcontracts.
  - Supplier Contracts.
  - Program Organization Diagrams.
  - Functions and Responsibilities of program teams.
  - Headcount Plan.
  - Product Integration and Test Plan.
  - Product Support Plan.
  - Program Safety Plan.
  - Quality Plan.
  - Sometimes others.

Chapter Twenty

Everyone Must Be Paranoid!

A frequent cause of a Program Turnaround Team not achieving their Turnaround Commitment is not being prepared for a new issue. These new issues can arise during the life of a program from a wide range of sources, including unexpected technical problems, unanticipated cost increases, changes in customer requirements, late or incomplete supplier deliveries, resignations by Program personnel, etc.

What is essential for a quick resolution of an issue is a "heads up" of the risk that it's coming (Figure 20.1). Nothing saves schedule time and general confusion more than having a problem resolution in hand when a problem occurs. So how do we do this?

- **A documented trait of all successful teams.**
- **A concern from any team member must be given due process.**
- **Awarded (do not punish) for identifying concerns.**
- **Process per Risk Management Plan.**
- **Share successes with program team!**

Figure 20.1 A Turnaround Program usually has many activities that can cause a risk or issue. It takes the whole Program Team to be on guard for unrecognized risks. Leadership must welcome all concerns.

20.1 Living Risk Management Plan

As discussed in an earlier chapter, the Program Turnaround must have a Risk Management Plan. This applies to any Program Turnaround, regardless of its size or content.

A Risk Management Team usually includes senior subject-matter experts who have the experience to know what problems may occur during the development within their areas of expertise. They are an excellent source both for providing that "heads up" of potential adversities and for assembling a well-thought-out risk avoidance plan before the issues occur.

It is important that this Risk Management Team stays vigilant and updates the description of Program risks and abatement progress as the Program matures. New risks may suddenly become apparent, and existing risks may never become issues. The plans for responding to identified risks and the issues that most likely would result must be current.

20.2 Team Members

Program leadership must pay very high attention to any concern about an unknown risk or issue brought forward by anyone on the Turnaround Team—not just by those directly on the Team, but also by the customer, the subcontractors and suppliers, and the enterprise. The door must be wide open to hear any concern from anyone about new risks and issues for the Program. The origins of this concern can be as simple as two team members discussing a worry they realize they share.

This is another reason it is so important for Turnaround leadership to have a real open-door policy. No team member should ever feel intimidated when visiting his or her immediate leader or the Turnaround Lead. Team members should be commended for bringing a potentially new problem or risk to the attention of leadership, even if has already been identified elsewhere.

Yes, occasionally a team member may share a worry about something that has an extremely low probability of happening or has no consequence. Some team members may even refer to this individual as a "worry wart." During the urgency of a program turnaround, this kind of input may sometimes seem like a distraction.

However, I highly recommend that leadership does not in any way imply that any concern brought forward is a distraction or does not merit consideration. Leadership must be always patient when any worry is brought to them. I have found that a real risk or issue is often hidden in a concern that initially may sound innocuous. I have encountered very few chronic "worry warts" on Turnaround Teams during my career.

All the members of the Turnaround Team, including subcontractors, suppliers, customer, and enterprise, must be regularly encouraged to be on the lookout for and report risks or issues that had not been identified yet. The Program will need time to plan a solution. As mentioned before, one of the worst impacts on program schedule and cost results from encountering a problem without any knowledge that it might occur or any solution planned.

20.3 Schedule Reviews

Sometimes an unforeseen issue in a program immerges very slowly. It may start to appear as an accumulation of small incidents, each one at first thought to be uncorrelated and insignificant. The most frequent source for discovering these kinds of problems are the reviews of program schedule performance.

An issue can become apparent when a development area or task is regularly slipping their commitment dates. This can indicate an erroneous prediction of schedule time required during planning of the work, but it can also indicate that a recurring problem or fault has crept into the Program. Examples include the use of development processes that are not as productive as anticipated or schedule bottlenecks resulting from insufficient headcount in some area.

This is why it is so important to update the completion schedules frequently—as a guideline, at least once every two weeks. These updates usually should be performed more frequently during intense development periods, making it easier to identify a task or work area that is encountering more than expected problems. This allows the Program time to solve developing issues before they become big ones.

There is an important benefit from having a professional scheduler looking for risky trends in the program schedule performance. It is uncanny how a good scheduler can observe a risky trend or other subtle anomaly in schedule performance, often before the Program leadership does. This early detection can give the Program more time to solve the developing problem before it becomes a major issue. A good scheduler will often save the Program cost, schedule time, and sometimes even embarrassment.

20.4 Schedule Reserves

Using schedule reserves at the Turnaround Lead level is the last resort to lessen or negate the effect an unanticipated issue, but the Lead may have to use their reserves to solve a problem that had not been anticipated.

Issues that were not anticipated are usually more expensive to resolve, because these kinds of problems most often do not give the Turnaround Team the time to either mitigate the risk leading to the issue or to develop a resource-efficient, low-impact response to the issue before it happens. Each additional day the Program Team must be funded to achieve the Turnaround Commitment beyond the scheduled completion date creates more expense.

All early warnings of risks to the Program can keep the Turnaround Lead and task leads from having to dip into schedule reserves. These warnings from any Program Team member are of great benefit for the Turnaround Program to help ensure achieving all promised completion dates.

20.5 Chapter Highlights

- An active Risk Management Plan is an important source of vigilance.
- A "paranoid" team is a valuable attribute for a Turnaround Program
- Evaluate and record risks brought forward by all team members.
  - Examples of team members identifying risks that are used to avoid issues should be shared with the Program Team.
  - Risk identification should be encouraged. Consider incentives.
- Review Program schedule status frequently to uncover subtle risks.
- Program schedules must be reviewed and updated when risks become issues.

Chapter Twenty-One

Team Dedication and Mentoring

The Turnaround Lead and the Program Turnaround leadership must manage an inherent dichotomy when working with the Program Team.

On one hand, as the Turnaround Lead, you must refrain from developing close personal relationships with the team members. This is because you may have to start the Turnaround by making some radical changes in the Program, and you may have more such changes to make in the future. You are also establishing a new program pace. With this, you cannot allow the perception of special personal ties between you and any of the Turnaround Team members. You must be perceived as impartial, as fair. The perception of being fair is essential to leading a strong and dynamic team.

On the other hand, you want to reach into the passions of each team member and garner their awareness and satisfaction of doing what some thought could not be done. You want them to savor the fact that they are growing professionally from this experience. You want them to reach for higher commitments than they have ever made before, and then give them a second chance if they fail. You want them to wake up in the morning looking forward to the challenges of the workday, even while knowing they may never have worked so intensely or with so much uncertainty about what they need to do to best serve the team. You want them to look back ten years from now and think that this was the most satisfying program they ever worked on and that they would do it again. This kind of challenge may not be what all enterprise employees wish to be involved in, but those supporting a Program Turnaround must know they do.

So how do you personally motivate your team yet avoid personal allegiances? The team must know you care for every team member the same. One of the best ways is to be a mentor.

Some Program Team members are not interested in being mentored. They may just want to do their jobs the best they can per the guidance they have derived from the professional experiences they have had during their career. You will still have their backs, like you will every other team member, but avoid offering unsolicited guidance other than by your example.

Others may feel they can collect some valuable career lessons from your example, your leadership of the team, and work-related dialogue with you. In these cases, you can be a mentor—a valuable source that identifies their professional strengths, answers their questions, enhances the lessons learned from mistakes, and even sets the bar higher for them so they can grow. As a mentor, you can do all this without being their "buddy."

21.1 A Real Open Door

How many times have you heard the term, "open-door policy"? I am reluctant to use it because it is such a cliché. When executed completely, it is a powerful symbol and icebreaker, but I believe leaders often do not fully subscribe to what this phrase really means.

On one program, I literally removed my office door. I laid it out in the entranceway so everyone could see that I meant it. The message was simple:

- Anyone can walk in and start a dialogue at any time. If I'm busy, I'll promise to get back to them that day, even if I do so that evening or via cell from the airport.
- Second, there will never be a negative outcome from talking to me, even if you come to tell me you think my conclusions or my approach are wrong. I will ask you why you believe what you do and listen to your answer very carefully. There may well be something I hadn't seen. At worst, we'll agree to disagree. I will continue to treat you the same as every other team member and will constantly be on the lookout for any growth opportunities I can offer. This is my promise. I encourage comment from all team members.

21.2 Triggering Strong Dedication

Dedication to enterprise or program seems to be out of vogue these days. Some leaders have suggested that in our current U.S. economy, we should value ourselves as professionals who perform a specific function in an efficient way while staying up to date.

Some propose that this new breed of professionals should be ready to plug into a project or program that needs their particular ability and unplug when they are done. Members of this new breed to some extent become itinerate workers, keeping themselves employed with the growing accomplishments they gather in their résumé and the networking they have established. The employee becomes solely responsible for deciding what they are good at and what additional experiences they need to get better. They take full responsibility to find the opportunities to receive the professional schooling and experiences they want.

This approach is great if you want, as a program lead, to get just the work you paid for. Many managers hire independent consultants to do this work. But for a Turnaround Program and most other development programs, leadership must strive to provide more than just pay for the work the Program Team members perform; they must offer an environment that engenders passion in the team members for their contribution and their future. Remember, we are human beings. We each require a complex and unique formula to be our best.

I would like to make some recommendations for getting "more than you pay for" as a program leader.

21.2.1 When It's Scary, Be Strong

This is a simple lesson that many readers may know, but it bears repeating. When the Program encounters what may appear to be an impossible problem, to whom does the Program Team look to evaluate how serious the problem is? The Turnaround Lead and their leadership. They not only listen to what they say, they also examine their body language. Any appearance of panic or bewilderment will not only imply that the Lead may not know what to do next or, even more serious, that the Lead thinks the Turnaround Team may not be able to surmount the problem.

This second implication can trigger a chain reaction of withdrawal from the Program by all team members. It can devastate a program and may leave at least a suspicion that the Turnaround Lead does not have complete faith in their ability.

It is very important to remind leadership that they are always "on camera." Their facial expressions and body language can express more than their words. When an unexpected issue hits the Program, the leadership team must stop, smile, be calm, and believe in the strength of their teams.

If you, as a lead, find yourself at a loss for words and/or unable to provide a direction on what to do next, it is likely because you and the team do not have enough facts to fully evaluate the situation. The best direction to give is to gather more information about the problem. Maintain your composure and remind the team by your example that they work best with complete evidence and facts. The Turnaround Lead must always leave the unquestionable impression that the Program is moving forward—despair and futility must not be in the leader's emotional vocabulary.

In addition, the message must be sent that the team must fully understand the new problem before determining a solution. The Lead must steadfastly believe and express this conviction, despite the panic that may be brewing within the team. Leadership must present utter confidence that the team can successfully solve the problem. With this conviction from leadership, they will.

21.2.2 Have Their Backs Even When They're Scared

A corollary of the above solution exists when an unexpected and seemingly overwhelming problem occurs in the work being performed by an individual team member. The Turnaround Lead must be completely calm and show high confidence that this team member will solve the issue. The Lead must give that team member time to fully understand the problem. If this team member is relatively new in their specialty, the Lead might suggest that a more senior team member tag along to provide second opinions and help in other ways. But the Lead must make it clear they are not taking the responsibility for solving the problem away from this less experienced team member—they still must carry the solution across the finish line.

Standing behind an individual contributor this way is a win for everyone and for the Program. I was treated this way decades ago as a struggling engineer trying to show the world I could "make it better." I recall the warmth of relief and gratitude that I felt when I was assigned a highly respected senior "wing man" to work with me on my problem-solving logic. I finally found the root cause of a very subtle problem in an important control system. I recall how both our mouths dropped when I finally saw it. But I never would have found it without this person's supportive and brilliant guidance. My gratitude for his help and the support of the enterprise will always remain with me.

21.2.3 Everyone Gets a Second Chance

When I was seventeen, I worked part-time as a ramp service person for an airline on the West Coast. OK, my role was also referred to as a "bag man." I was one of the ground crew that unloaded the arriving passenger's baggage, and then loaded the departing passenger's baggage before the plane left for its next destination. We drove tugs that pulled the baggage carts between the airplanes and the airport facility.

As new employee, I wanted to show I could drive the tug and empty the carts quickly. But as I drove away from the airplane with carts in tow, some of the luggage fell onto the asphalt ramp. I could hear a heart-rending "aaaaaaaaaaaaaah" from the scores of people on the airport observation deck watching me. I quickly stopped and ran to place the fallen luggage back on the carts when two suitcases opened up! Cloths, toiletries, and underwear were now all over the ramp. This time a much louder "AAAAAAAAAAAAAH" came from the observation deck. I hastily stuffed these items back into the suitcases, put them on the cart, and drove them slowly to the passenger pickup area with my head down.

I remember thinking it could not be any worse. I had made a complete fool of myself in front of scores of airline passengers, the flight crew, and fellow ground crew workers. I was certain my job was toast!

The station chief for the airline called me into his office that day. He was a very polished leader and respected by everyone. I knew I had blown it and assumed that I was going to be fired. I figured, sometimes you try your very best but you just don't have what it takes for the job.

The young, clean-cut station chief sat across from me and simply said, "I heard you had some problems on the ramp today. I give everyone a second chance. Don't let it happen again."

"What?" I thought! "You're not going to fire me? A second chance? Despite my feeling like the lowest form of life at the airport, you're on my side?"

My gratitude exploded in me like a bomb, not so much because I still had a job, but because I felt someone appreciated that I was trying so hard to do my best. I recall reviewing every moment of my embarrassing baggage incident in my mind and inserting checks in my work that would never allow such a thing to happen again. It never did!

As a leader in a program, you can divide the mistakes made by your reports into two causes. One is that the employee may not like their work, may not care to do their best, and may have made the mistake because they did not care enough to avoid it. The second is that the employee is trying with all their heart to do the best job they can, but did not have the experience or knowledge to avoid this mistake.

In the first case, there may be a bad work attitude or lack of motivation that should be discussed with the employee. The enterprise, the Program, and this employee may be better off if the employee is offered another assignment.

But if the employee's mistake falls into the second category, they should be given a second chance. Note, it's a "second chance," not "another chance." The employee cannot be left with the impression that their mistakes will be tolerated indefinitely.

If the employee really wants to do the best they can, they will thoroughly scrutinize their mistake and make sure it will never happen again. They may ask you for some insight and/or guidance when fixing the cause of their mistake, but they will take responsibility for doing the job correctly in the future.

This approach applies not only to inexperienced new employees but also to senior employees starting new tasks. For example, an excellent senior technical staff employee may want to try taking on more leadership responsibility. They may make some basic leadership mistakes because they never had this role before. This new leader's management may be tempted to remove them from any future leadership roles because they feel this person is just not a "natural leader." But good leadership ability is mostly learned and not inherent. This senior employee may never have been given any basic leadership opportunity or training, but they can learn it if they are given a second chance.

21.3 Stand Up for Them When There Is a Special Problem

The Turnaround Program leaders must ask for more dedication, innovation, and hard work from their team members than they may ever have provided before. Yet, on the other hand, the Turnaround Lead and Program leadership team must vigorously protect their team members when they are being challenged. Some examples of these challenges follow.

21.3.1 Outside Criticism

Sometimes a Turnaround Team will receive criticism from organizations outside of the Program, whether from other programs in the enterprise or from other companies or businesses outside of the enterprise. This sometimes stems from jealousy or disbelief that the Program Turnaround is doing well. The unsubstantiated accusations may include that the Turnaround Program is publishing false metrics, the Program is getting special help from the enterprise, or the Program stole all the best people from the other programs.

Program leadership must respond to these claims. The first thing leadership must do is determine if there is any truth to them. If there is, leadership must apologize for any imposition on anyone and announce a plan to remedy the problem, then periodically review the progress of the remedy to closure. The source of the complaint as well as the Program Turnaround personnel will respect the openness and integrity of Program leadership for responding this way.

If the criticism has no basis, leadership must issue an immediate response with all facts that counter the accusation. If the criticism is repeated and there is no new information brought forward to support it, it should be ignored.

What is very important is the high amount of conviction and strength the Turnaround Lead and Program leadership demonstrate while refuting any malicious claims.

21.3.2 Poor Work Performance

Sometimes the leadership for a Turnaround Program may find themselves in a situation in which they want to help a team member to correct their shortcomings, but this team member continues to repeatedly fall short of delivering a quality result on time.

What is important is to treat this situation with respect and patience. The team member may not have the experience or training to perform their assigned work to the expectations of the leadership. Instead of just removing the individual from the role, Program leadership should devise a plan to improve the skills of this team member. Their plan may require that the team member spend extra hours for special training and/or remedial assignments. If the team member is motivated to find a way to improve their contribution, they will make the sacrifices. The resulting allegiance a team member will have to leadership and the Program for helping them become an excellent contributor will then be very high.

If, on the other hand, the team member is not interested in working to improve their performance, this should be obvious to both the team member and leadership. Both should understand that the team member is not doing well executing the assigned work, and it should be no surprise to the team member when they are given a new assignment in or out of the Program.

If this delicate situation is handled correctly by leadership, this team member should leave the Program with no resentment against the Program or enterprise. This team member should respect leadership for deriving and trying a plan to keep them on the Program.

The dedication of the rest of the Program Team will increase when they see leadership trying to improve the performance of a fellow team member.

21.3.3 Reduction in Workforce

I worked on a program during my early career that grew to over two thousand people. It was clearly on step, it was very successful, but it was cancelled with short notice.

Those executives that take pride in "making the hard decisions" might have given each team member a fixed duration to contact the members in their network and hopefully find an assignment on another program. Despite the past dedication and hard work of each team member, this approach would be to assign them the responsibility to find their next position. If they had not found something by the time their search duration was exceeded, they would be laid off.

But for the program I was on, the president of the enterprise mandated that the managers of all the other enterprise programs find these people jobs on their programs. Not just any job, but jobs that applied the talents these people had developed in the program that was cancelled. For many workers, the cancellation actually resulted in a step up in their profession (I benefited this way).

This solution may have resulted in a slight and temporary reduction in business performance for some of the programs. After all, some of the programs may have ended up having more employees than they needed for a while. But the increase in enterprise dedication garnered by this action, not only in the hearts of the people on the doomed program but also for all the people on the other programs that witnessed this, was tremendously high and lasting. Not to mention the benefits of the new perspectives and innovations brought into some of the programs absorbing these talented orphans.

This was truly wisdom from the leader of our enterprise. The solution had a much greater long-term benefit to the competitiveness and productivity of the enterprise than the short-term cost savings the alternative response would have brought.

I recommend the Turnaround Lead think well beyond the quarterly numbers when dealing with a reduction of force. The trust and dedication of the enterprise employees are a commodity that an enterprise must preserve to be the best. An employee will not reach into their deepest corners and give their very best to a program if they don't feel that their well-being is a top priority in the eyes of the Program and the enterprise.

The good Turnaround Lead and enterprise executives will know that enduring a short period of reduced business performance to support the job stability of these employees is a small price to pay to maintain their dedication and best efforts. The displacement of poor performers will then naturally occur in the programs with excessive headcount. This process preserves the best talent and hardest workers for the enterprise. Long-term employment with the enterprise should be based on the work performance of each employee, not the circumstances of the business, which the employees cannot control.

21.3.4 Personal Crisis

As discussed earlier, any team member supporting the Program Turnaround must have the ability to "pull the line" (that is, step away temporarily from the

Program) if they have a personal crisis. If this happens, leadership should support the resolution of the issue, within reason.

Here again, leadership must walk that fine line between being sensitive to and supportive of the needs of the individual contributors on the one hand, and not developing a personal bond with any member of the team on the other. It is not the role of the Lead to solve team members' personal problems, but instead to give them the freedom and any support possible to solve their issue. The team member must be reassured that the Program and enterprise will do all they can to help preserve their safety and happiness without the threat of demerit or retribution.

In our modern world, the reasons for pulling the line can be quite varied. They can include a family health crisis, child behavior issues, child rearing support, loss of a family member, a major house maintenance issue, work burnout, etc. What is important to a team member should be important to the Program.

Current leaders reading this may understandably think that this approach to a team member's crisis opens the door for abuse, such as employees taking days off from work with few or no real issues to solve. It has been my experience that if the Program Turnaround comprises excellent team members, the opposite occurs. In fact, members of good turnaround teams often do not take all the personal time off they probably should.

It is an obligation of good leadership to keep a sensitive eye on all team members. Leadership must look for individuals that may be showing signs of not being able to perform their best due to some outside distraction. Give this person some room to solve their issue. Respect and maintain their privacy.

Again, it is not leadership's role to solve the issues in the personal lives of the team members. But the members of the Program Team must know that without question they will be given the time and any other support the Program can provide to solve big issues in their lives.

21.4 Seeing in Them What They Don't See

A leader often has as a few advantages that can be of great value to their reports. First, they can evaluate these employees from a different viewpoint. This may provide some additional insight the employee does not see about what they are capable of doing.

Second, the leader usually has a broader perspective on what capabilities the Program and the enterprise need. They may be able to connect the employee to opportunities within the enterprise the employee was not aware of.

From their perspective, leadership may see potential in a team member to contribute even more to the enterprise than they are in their current role. This team member may not have been aware of this potential (see Figure 21.1). Trying

- **Identify the special talents your team members may not see.**
- **Give team members assignments that develop their talents.**
- **Advertise these potential talents to program and enterprise.**
- **This may identify latent problem solvers and leaders.**
- **Intense Turnaround pace often makes these talents apparent.**

Figure 21.1 As a leader, look for unharvested talents in those reporting to you. You both may be surprised.

to take action to apply an underused capability can be one of the most valuable gifts an employee can receive from their leader. It demonstrates the genuine care the Program and enterprise leadership have for the employee. It will help teach the employee more about their capabilities and perhaps develop their interest in a new career area—it can favorably change the course of the employee's career.

Encouraging a team member to explore an unrecognized talent should be easy. It can be as simple as, for example, "Kathy, you really understand package B, and folks respect your knowledge and the direction you've suggested. I'd like you to try leading the package B integration team." For some reason, Kathy may never have imagined herself leading a team; she may have just been trying as hard as she could to make package B as successful as possible. Now she realizes that someone she works for and respects her good performance is telling her she may be able to lead some of the package B effort. She realizes she may be able to make a bigger contribution to the success of package B as a leader than by just her individual contribution.

What is very important is that Kathy has come to this realization and was given the leadership opportunity because of the oversight and initiative of her leader. This opportunity probably never would have happened otherwise. She appreciates this extra effort by her leader, and her dedication to giving her very best to the Program and the enterprise is further enhanced by this special effort.

However, I must make a simple yet important point. The work to promote the special capabilities of their reports should not be a "special effort" for a leader. Looking for and tapping the career growth potentials of all their reports must be part of the daily work of a program leader. It develops the team's capabilities and helps instill high devotion from the employees.

Identifying a potential growth path for an employee is only the first step the leader provides team members. The leader must then offer assignments that allow the suspected additional talent to develop.

Often, special training is needed for the employee to take on the new suggested role. The leader should help identify this training and identify any assistance the enterprise may offer. In addition, the leader may want to assign a mentor (or be a mentor) to the employee as they develop in their new career direction.

Sometimes the employee and the leader will discover the employee does not exhibit all the work performance ability necessary in their new role. Or sometimes the employee will not want to take this new career path, even though they may have the inherent talent to do so. In cases such as this, the leader should understand that the growth opportunities they provide their team members might not always be successful. If the attempt fails, the employee will still gain additional insight into both their capabilities and the limits of their interests. However, if it succeeds, greater value will be added to the employee's future, to the Program, and to the enterprise.

21.4 Chapter Highlights

- Establish a real open-door policy.
  - Always be available.
  - Be a nonjudgmental confidant.
  - Make your policy visible.
- A path to strong employee dedication.
  - Have their backs when adversities occur within the enterprise.
  - Everyone gets a second chance.
  - Support and protect team members when they encounter problems outside of work.
- See what they don't see.
  - Suggest talents in a report that they may not be aware of.
  - Make assignments that exercise these talents.

Chapter Twenty-Two

Benefits for the Enterprise

It is understandable that the need for a turnaround program may be construed as a source of embarrassment for an enterprise. Some enterprise leadership may feel that having to resort to a new Program Plan is a serious failure in the way they set up the original program. However, as discussed at the beginning of this book, it is all too often easy to make the mistakes that can lead to a poor plan or the wrong leadership for a development program.

The fact is, an enterprise is really evaluated more by how quickly they address and resolve issues than by how they got into them. Moreover, there are real benefits to an enterprise from conducting a program turnaround.

22.1 Encourages a Culture of Achieving Program Commitments

Having the enterprise conduct a program turnaround will set an example and be an inspiration to the other enterprise program teams. It will clarify the necessity of always speaking in terms of "commitments," not just "goals," to the customer, to the enterprise as a whole, and to the other team members, and it can show how the Program Turnaround makes quantitative commitments in the four most important program performance areas—providing the required product performances, meeting promised schedule dates, staying within planned program cost, and managing risk. It can demonstrate how team members achieve their commitments with teamwork, innovation, and extra work hours, if needed. It demonstrates a process for quickly recording and understanding program development anomalies and failures and coming forth with a complete solution on the first attempt.

The other programs witnessing the work of the program save may be performing acceptably, but seeing the success of the turnaround work may deepen their pride in working for the enterprise, as well as improving their business performance and their morale. It will demonstrate the high value of achieving realistic promises and not just stating hopes.

22.2 Gives Employees of the Enterprise the Knowledge and Feel of a Team On Step

As discussed in the introduction to this book, a development program team, regardless of what they are developing, discretely transitions from a state of laboriously plowing through the daily tasks to being on step. Like a motorboat, when on step, a program's rate of progress achieves much more speed with just a little more effort.

When a program is on step, the Program Team is excited about the assignment of a clear and single owner for each task. They are delighted when each team member is given full authority to complete their assignment. They are gratified that management supports their high work efficiency and career goals. They are further grateful that management is making the extra effort to evaluate what extraordinary talents they may have as individuals.

Members of the Program Turnaround Team take tremendous pride in their ability to quickly solve problems, to move ahead on the Program Plan at unprecedented speeds, and to successfully achieve the Turnaround Commitment.

Team members of other programs in the enterprise will observe these positive elements of the Program Turnaround. Some may wish to participate in this hard-working but highly gratifying effort. They will see firsthand what a program on step looks like and how it might feel to be part of this kind of team.

The satisfaction of those who participate in the Program Turnaround can be contagious. Their exposure to team members of the other enterprise programs can prepare them to support new program saves if the need arises. It may even provide the inspiration for these team members to help move their home program to be on step.

22.3 Identifies the "Solvers"

We have discussed the difference between employees who are content to continue to *work on* a problem versus those who are motivated to *solve* it. The "solvers" motivate complete task closure by the team and often are promoted to become task leaders.

A Program Turnaround environment can quickly identify the team members who are "solvers" from those who are comfortable just working on a problem. Both types of professionals are important, but a Program Turnaround needs team initiative and innovation to achieve their Turnaround Commitment quickly. The Program cannot rely just on leadership to instill these qualities. Team members who have a high propensity to solve and close the development challenges are a necessary ingredient for a program turnaround to be successful.

The Program Turnaround can help Program and enterprise management to accurately identify the "solvers" and help them develop their leadership skills, if they wish to do so. Enterprise leadership can then assign these professionals to lead and motivate closure for teams in future programs.

22.4 Identifies Future Leadership

The best leaders for an enterprise are those who concentrate on its products and its business successes. For a successful enterprise, these attributes can be more important than the worker's tenure, work style, or breadth of work contacts. It is hard to trump the success of leading the delivery of contract-compliant goods with high quality, on or before the committed delivery dates.

In the dynamic, fast-paced work environment of a turnaround program, the ability to lead teams to deliver successful content on time will be more evident than in a program under less pressure. Enterprise management should take note of which leaders are succeeding in this environment, as they will likely be excellent choices to lead future enterprise programs. These leaders should be the first people considered for additional leadership training and promotion into higher levels of management.

22.5 Provides Process Improvements and Innovations for Future Programs

As discussed earlier, process improvements and product innovations are solicited from individuals and teams during the Turnaround. In addition to these improvements and innovations frequently becoming necessary for achieving the current Turnaround Commitment, they will be of great benefit to future enterprise programs. They should be reviewed by the appropriate enterprise organizations to determine if they should be added to the basic enterprises processes and/or standards.

22.6 Helps Prevent Mistakes in Future Programs

I have discussed some of the frequent mistakes that result in a new development program falling into trouble. As I mentioned, these mistakes often result from innocent assumptions. They become easier to identify in future programs after negative or harmful outcomes from them have been experienced.

Having to resort to a turnaround effort and a new program plan (Turnaround Plan) is one result of this bad experience—but it becomes an opportunity for the enterprise to review what went wrong, so it can be avoided in the future.

This review must not be intended to find a group or an individual to blame! No one deliberately organizes a program to fall into trouble, and this is a valuable opportunity to identify in detail why the program was starting to fail. At this finer level of detail, errors that are unique to the guidelines, processes, and/or organization of the enterprise will be identified. Including the root cause process used for test/development failures, discussed earlier, will certainly make this review more valuable, and the results should be used to amend the enterprise process, procedures, command media, or equivalent so the problem(s) do not occur again.

22.7 Increases Enterprise Morale and Allegiance by Showing How Successful They Can Be

A well-run enterprise is like a family—when a program is not doing well, the other programs in the enterprise become concerned.

When a troubled program successfully executes the new Program Plan and achieves its Turnaround Commitment, the other enterprise programs will share in the satisfaction of recovering from a problem. Fears that the enterprise may be in trouble dissipate. Team members from all the enterprise programs take pride in the troubled program's promptly determining that it had to take major action and succeeded in doing so. As stated before, program mistakes are to be avoided, but a more important measure of an organization is the speed and completeness of its recovery.

The enterprise should be proud of successfully achieving closure with a Turnaround Plan resulting in a program save. Within any data sensitivity or ownership limits, the enterprise should internally share that one of its programs was successfully put back on its feet!

22.8 Demonstrates High Capability of Enterprise Brand to Business Community

Of course, any publicity about the success of an enterprise program must be constrained to data sharing limits imposed by the customer and sometimes even the enterprise. However, the successful Program Turnaround should be shared with competitors and potential customers of the enterprise if possible. Having identified unacceptable program performance, then decisively managing a successful recovery, is a major plus for the enterprise brand. It shows that the enterprise promptly recognizes a program in trouble and manages a prompt and completely effective resolution.

22.9 Chapter Highlights

- Demonstrates the value of promising and achieving commitments.
- Identifies latent leadership.
- Provides insight to enterprise programs of repeated mistakes that should be avoided.
- Gives team members the feel of their team being on step.
- Increases morale and allegiance by demonstrating high enterprise success.
- Identifies good problem solvers.
- Creates valuable innovations that can be used in future programs.
- Fortifies enterprise enthusiasm for implementing continuous work improvements.
- Demonstrates high capability of the enterprise brand to the business community.

Index

D

E

G

H

I

N

O

P

Q

R

S

U

V

W